Handbook of Image-Based Security Techniques

Handbook of Image-Based Security Techniques

Shivendra Shivani
Suneeta Agarwal
Jasjit S. Suri

CRC Press
Taylor & Francis Group
Boca Raton London New York

CRC Press is an imprint of the
Taylor & Francis Group, an **informa** business

CRC Press
Taylor & Francis Group
6000 Broken Sound Parkway NW, Suite 300
Boca Raton, FL 33487-2742

© 2018 by Taylor & Francis Group, LLC
CRC Press is an imprint of Taylor & Francis Group, an Informa business

No claim to original U.S. Government works

Printed on acid-free paper
Version Date: 20180427

International Standard Book Number-13: 978-1-1-38-05421-9 (Hardback)

To my beloved family members for encouraging and admiring me to believe that I am capable of authoring a book.
Shivendra Shivani

To my husband for his untiring effort and support.
Suneeta Agarwal

To all my collaborators around the world.
Jasjit S. Suri

Contents

Section II　**Digital Image Watermarking**

Chapter　5 ▪ Digital Image Watermarking: Introduction　145

CHAPTER 10 ▪ Dual Watermarking   293

SECTION III Steganography

CHAPTER 11 ▪ Steganography   323

Foreword

With the rapid advancements in technology and the popularity of the Internet, illegal modifications in digital media have become easy and difficult to prevent. Therefore the protection of digital media and its property rights have become vital issues of concern. Protecting digital data is extremely important and an emerging area of research. Many efforts have been made in this area by the cryptographic community. Images are treated as very powerful digital media which must be protected on the web. There are three image-based security mechanisms which reported till now are Visual cryptography, Digital Image Watermarking and Steganography. Specifically, visual cryptography allows effective and efficient secret sharing of images among a number of trusted parties. Visual cryptography provides a very powerful technique by which one secret image can be distributed into two or more shares. When the shares, printed onto transparencies are superimposed together, the original secret image can be recovered without any computation. While digital watermarking and steganography seem very closely related terms, they are actually very different in nature and practice. Watermarking is used to protect the ownership information or content of the original cover image with or without tamper recovery capabilities whereas steganography is used to protect any secret image or text behind the image.

Preface

This book is especially written for young researchers who want to take their research direction toward image-based security techniques. This book is written from the beginner's perspective with the commitment to make them fully capable of understanding all latest research gaps in this field. Our goal with this book is to provide a framework in which one can conduct research and development of image based security technology. This book is not intended as a comprehensive survey of the field of visual cryptography, watermarking and steganography. Rather, it represents our own point of view and our own technical contribution on the subject. Although we analyze specific examples from the literature, we do so only to the extent that they highlight particular concepts being discussed.

PURPOSE

Our purpose with this book is to provide guidelines to readers which are used to enhance the research-oriented concepts of visual cryptography, watermarking and steganography. This book represents ground truth as well as our own perspectives on this subject. Some well-known examples are taken from the literature to highlight the particular concepts. We are no exception, our own backgrounds being predominantly in images and security. The fundamental principles behind all the security dimensions of images are same, so we have made an effort to keep our discussion of these principles generic. This book is easily accessible to those who do not know much about images and security fundamentals. In one sentence, we can say that this book offers encouragement and empowers readers to rejuvenate their innovative imaginations in the field of image-based security.

CONTENT AND ORGANIZATION

This book is divided into four sections. The first section includes Chapters 1,2,3 and 4 which deals with the concept of visual cryptography from the basic to advanced level. The objective of this section is to make researchers aware of this latest technology of secret sharing. Each chapter in this section provides the fundamentals for the visual cryptography under consideration and a detailed description of the relevant methods, presenting examples and practical applications to demonstrate the functioning of visual cryptography.

Chapter 1 provides introductory material for visual cryptography, preliminaries, various evaluation parameters as well as a wide variety of applications of VC that serves as motivation. The applications highlight a variety of sometimes conflicting requirements for VC, which are discussed in more detail in the second half of the chapter.

Chapter 2 deals with the broad classification of VC on the basis of various applications and subjective evaluation parameters.

Chapter 3 perfectly deals with all classes of VC like Halftone Visual Cryptography (HVC), Progressive Visual Cryptography (PVC) etc which are actually made for halftone images as input which require no computation at all at the recovery end.

At the end of this section **Chapter 4** deals with all classes of VC like Multitone Visual Cryptography (MVC), XOR-based Visual Cryptography, Multi Secret Sharing (MSS) etc which are actually made for perfect recovery and require computation at the recovery end.

After reading this section, one will be able to think about the research gaps in the field of VC and be able to customize the existing research.

The second section includes Chapters 5,6,7 8,9 and 10 which deals with the concept of Digital Image Watermarking from the basic to advanced level. After reading this section, one will be able to judge the better watermarking algorithms, able to develop his own algorithm and customize all existing algorithms in an efficient manner so that they give more effective results.

Chapter 5 provides introductory material on watermarking. It provides a history of watermarking, as well as a wide variety of applications of digital watermarking that serves as motivation. The applications highlight a variety of sometimes conflicting requirements for watermarking, which are discussed in more detail in the second half of the chapter. This chapter also presents several frameworks for modeling watermarking systems. Along the way, we describe, test, and analyse some simple image watermarking algorithms that illustrate the concepts being discussed. This chapter analyzes message errors, false positives, and false negatives that may occur in watermarking systems during the evaluation of watermarking techniques.

Chapter 6 deals with one of the most important classes of watermarking which is generally used in content authentication. Here we deal with various state-of-art approaches for block-based and pixel-based fragile watermarking techniques in the environment of various intentional and unintentional attacks. Here we also deal with the recovery techniques of the removal for the image content.

Chapter 7 deals with more powerful fragile watermarking techniques having the capabilities of tamper detection as well as recovery without intervention in the original content. This facility is provided in the spatial domain of the image.

Chapter 8 deals with more powerful fragile watermarking techniques having the capabilities of tamper detection as well as recovery without intervention in the original content. This facility is provided in the frequency domain of the image.

Chapter 9 deals with one of the most important classes of watermarking which is generally used in copyright protection. Here we deal with various state-of-art approaches for robust watermarking and their behaviour in the presence of noisy environments.

Chapter 10 provides an efficient approach toward the making of a special kind of watermark having the qualities of being both a fragile as well as robust watermark.

The third section includes Chapter 11 and Chapter 12 which deal with the concept of steganography and steganalysis. Here we deal with the notion, terminology and building blocks of steganographic communication. In **Chapter 11** we deal with the basic terminologies and understanding of steganography whereas **Chapter 12** provides some practical approach towards steganography and steganalysis. The fourth section contains somewhat advanced research topics in this field. This section contains two chapters: Chapters 13 and 14. Here **Chapter 13** deals with all types of hybrid approaches made by two or more image-based security techniques and **Chapter 14** deals with protection of other multimedia objects like video, audio, etc.

AUDIENCE

Primary: This is a complete textbook on image-based security which deals with all three security dimensions of images. Due to its excellent introductory content and motivations in the form of various applications, this book may be treated as a startup book. Hence this will help all those novice readers who want to start their research in the field of image based security.

Secondary: Finding novel problems is more difficult than finding their solutions. For the same reason, this book also deals with all potential research gaps which may be treated as a future scope of research in the field of image based security. Hence this book is again most appropriate for those researchers who are already involved in this field but trapped somewhere because they do not have proper problem statements.

UNIQUE ANGLES

In this book, the authors:

1. Present the implementation of each concept used in this book in MATLAB® as a whole.

2. Present the stepwise functions of MATLAB® wherever required in between the text, so that the reader can understand the theoretical and practical concepts of image-based security simultaneously.

3. Present many encouraging real-life examples and applications which will definitely motivate the users.

READER BENEFITS

1. Novice readers will enhance their basic understanding of security and images.

2. Readers will learn how to start thinking in research directions on image-based security.

3. Researchers will see new domains for research.

4. Readers will be able to implement the concept of security on images using MATLAB®.

5. Readers will also be able to develop novel real-life applications where images are used for authentication in various ways.

6. All chapters of a section are well connected. A reader can start any section because the sections are not dependent on each other.

ACKNOWLEDGEMENTS

My first debt of gratitude must go to my advisor, Prof. Suneeta Agarwal. I could not have imagined having a better advisor and mentor in my life. Whatever I say of her in this acknowledgement is not enough to describe her knowledge and skill. She patiently provided me the vision, encouragement and advise throughout the writing of this book.

I am also grateful to all members of my buddy group (Shailendra Tiwari, Ashutosh Agarwal, Raman Singh, Nitin Saxena, Vipin Pal and Vinay Gautam) with whom I was able to hold qualitative discussions and for pushing me hard in my free time to write this book. In the end, I am very thankful to my parents, Bhaiya, Bhabhi, Didi, Jija, younger brothers & sisters and my fiancee who dreamed of this time and instilled in me with dreams of research and its wonders, and watched with affection my journey towards their dream and my goal.

Shivendra Shivani   Suneeta Agarwal   Jasjit S. Suri

Authors

Dr. Shivendra Shivani received Ph.D and Master's degrees, in information security (digital image watermarking and visual cryptography) from the National Institute of Technology Allahabad, India. Currently he is working as Assistant Professor in the Department of Computer Science and Engineering of Thapar Institute of Engineering and Technology (TIET), Patiala, India. His research interests include digital image watermarking, visual cryptography and steganography, biometrics, security, gaming and animation, etc. Dr. Shivani has published various research articles in many reputed international journals and conferences in the field of visual cryptography and watermarking. He has also given expert talks in these fields many times. He is a member of various professional bodies including, Cryptology Research society of India (CRSI), IEEE etc.

Dr. Suneeta Agarwal received the B.Sc. degree in 1973 from the University of Allahabad, the M.Sc. degree in 1975 from the University of Allahabad, and a Ph.D. in 1980 from the Indian Institute of Technology Kanpur. She has more than 40 years of teaching experience and is currently a professor in the Computer Science and Engineering Department, National Institute of Technology Allahabad, India. Her current research interests include visual cryptography, 2D and 3D image watermarking, steganography, face recognition etc. She has guided more than 15 Ph.Ds and more than 100 M.Techs in these areas of security. She has published more than 250 research articles in the field of image security in various reputed international journals and conferences. She has also edited many book chapters.

Dr. Jasjit S. Suri, PhD, MBA, Fellow AIMBE is an innovator, visionary, scientist, and an internationally known world leader. Dr. Jasjit S. Suri received the Director General's Gold medal in 1980 and the Fellow of the American Institute of Medical and Biological Engineering, awarded by the National Academy of Sciences, Washington DC in 2004. He has published over 550 peer-reviewed articles and book chapters with H-index (40), and developed over 100 innovations/trademarks. He is currently Chairman of Global Biomedical Technologies, Inc., Roseville, CA, and on the board of Athero-Point, Roseville, CA, a company dedicated to Atherosclerosis Imaging for early screening of stroke and cardiovascular monitoring. He has held positions as chairman of the IEEE Denver section and an advisor board member to healthcare industries and several schools in the United States and abroad.

I

Visual Cryptography

Visual Cryptography: Introduction

CONTENTS

WITH the rapid advancements in technology and the popularity of the Internet, illegal modifications in digital media have become easy and so difficult to prevent. Therefore the protection of digital media and its property rights have become a vital issue of concern. Protection of digital data is an extremely important and an emerging area of research. Many efforts have been made in this area by the cryptographic community. One of the data security methods known as Visual Cryptography (VC) is dealt with in this chapter. Specifically, visual cryptography allows effective and efficient secret sharing among a number of trusted parties. Visual cryptography provides a very powerful technique by which one secret can be distributed into two or more shares. When the shares, printed onto transparencies are superimposed together, the original secret can be recovered without any computation.

The **Chapter Learning Outcomes (CLO)** of this chapter are given below. After reading this chapter, readers will be able to:

1. Understand the introduction and need for visual cryptography.

2. Understand the concept of the formation and validation of a basis matrix.

3. Understand evaluation parameters for visual cryptography.

4. Identify different dimensions of visual cryptography.

5. Identify various research gaps present in existing approaches to visual cryptography.

1.1 INTRODUCTION

Secret sharing has a vital cryptographic application. It is used in scenarios when people, involved in cryptographic processing of private data (secret), are either unreliable, or do not have trust in each other, while they together want to secretly compute some function for their private data. Hence to conceal a secret, it is split into various pieces called shares and distributed among participants, involved in cryptographic processing, in such a way that the secret can only be recovered by certain subsets of the shares. The search for efficient secret sharing schemes is still a great focus of research for cryptographic community.

The idea of secret sharing was first introduced by Shamir in 1979. This scheme was based on the use of Lagrange interpolation, polynomial and the intersection of affine hyperplanes. Since then, much research have been carried out for the suggestion of different efficient secret sharing schemes. Most of these schemes are based on different mathematical primitives, such as matrix theory and prime numbers.

Secret sharing approaches are specifically designed for numeric data and text. Multimedia data such as image, audio and video are distinctive in nature from numeric data and text. Since they have a huge amount of information and the difference between two neighboring values is typically very small, it is considered very difficult to apply traditional secret sharing schemes directly on multimedia data.

In the current era, images are considered more essential than other text

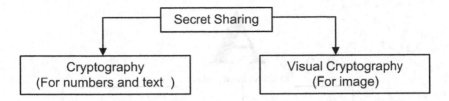

FIGURE 1.1 Classification of secret sharing.

sensitive information. With the rapid advancements in technology and popularity of the Internet, illegal modifications in digital images have become very easy and so difficult to prevent. Therefore the protection of digital images and their property rights have become a vital issue of concern. Visual secret sharing schemes where images are secrets, are a visual variant of the ordinary secret sharing schemes which are based on the human visual system. A visual categorization of secret sharing schemes can be seen in Figure 1.1.

Points worth remembering:
A secret may be either in the form of text, numeric or multimedia file.

1.2 VISUAL CRYPTOGRAPHY

Visual Cryptography (VC) is a type of secret sharing scheme proposed by Naor and Shamir in 1994. This cryptographic approach is especially meant for protecting the images which is why it is called visual cryptography as the word *visual* refers to the image. Visual cryptography provides a powerful technique by which one secret image can be divided into two or more shares. When a predefined set of these shares are superimposed exactly, the original secret image can be discovered without any computation otherwise nothing will be revealed. A k-out-of-n or (k, n) is a special type of VC scheme shown in Figure 1.2, where encoding of a secret image (binary in nature) into n visually random shares is illustrated.

After making transparencies of n generated shares, they are distributed among n participants, one to each. Any single share does not reveal any visual content of the secret image. But when k or more assigned transparencies of the participants are properly superimposed together, visual content of the secret image will be exposed. It means that at the receiver end, the secret decoding process requires no computation at all, i.e., the beauty of the visual cryptography. In spite of very high computation power, decoding of secret binary images is very tough with the help of only $k-1$ or fewer transparencies of the participants. To demonstrate the creation of shares and principles of VC, consider example 1.1 where a trivial codebook for 2-out-of-2 VC schemes is explained.

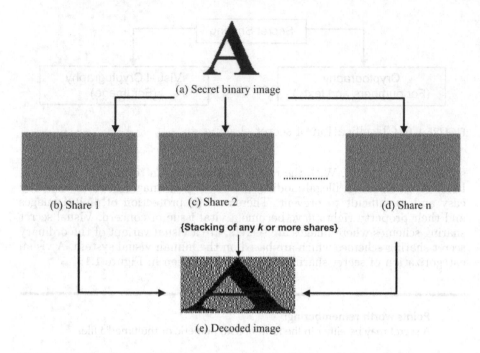

FIGURE 1.2 k-out-of-n VC, where a secret image is decoded by stacking k or more than k shares.

Points worth remembering:
Visual cryptography is a cryptographic scheme used to protect and secretly share the images.

Points worth remembering:
In traditional visual cryptography secret decryption is done with no computation.

Example 1.1: A simple 2-out-of-2 VC ($k = 2, n = 2$) scheme is shown in Figure 1.3 which is a special case of the k-out-of-n VC scheme. Each pixel p of a secret binary image is encoded into a pair of black and white subpixels for each of the two shares using the codebook shown in Figure 1.3. If p is white/black, one of the first/last two columns under the white/black pixel in Figure 1.3 is selected randomly so that the probability of selection will be 0.5. Thus values of the first row of it are assigned to the first share and values of the second row to the second share. Irrespective of whether p is black or white, a single pixel is encoded into two subpixels of black-white or white-black with equal probabilities. Thus an individual share has no clue about whether p was black or white. The last row of Figure 1.3 shows the stacking of the two shares. If the pixel p is black, the output of stacking will be two black subpixels. If p is white then the result of stacking will be one white and one black subpixel. Hence by stacking two shares together, one can obtain approximate visual information of the secret image. Figure 1.4 shows an example of the 2-out-of-2 VC scheme.

Pixel				
Probability	50%	50%	50%	50%
Share 1	▮	▮	▮	▮
Share 2	▮	▮	▮	▮
Stack Share 1 & 2	▮	▮	▬	▬

FIGURE 1.3 Codebook of 2-out of 2 VC, where a secret pixel is encoded into two subpixels in each of the two shares.

Figure 1.4(a) shows a secret binary image I_{sec} to be encoded. According to the encoding codebook shown in Figure 1.3, each pixel p of I_{sec} is divided into two subpixels in both the shares, as shown in Figure 1.4(b) and (c). Stacking these shares leads to the output image shown in Figure 1.4(d). The decoded secret image is clearly revealed. The width of the reconstructed image is just twice of the original secret image and there is some loss of contrast.

1.3 APPLICATIONS OF VISUAL CRYPTOGRAPHY

Visual cryptography can be used for enhancing security for many practical applications in real life. Various proposals have been suggested by researchers for applications of visual cryptography. Some have been implemented and some are yet to be implemented. Here we discuss a few of them.

1.3.1 Trojan-Free Secure Transaction

A very interesting application of VC is Trojan-free secure transactions. Nowadays due to the popularity of the Internet, usage of online transactions is increasing at a burgeoning rate. Online transactions are more vulnerable to man in the middle attacks than offline transactions. Trojans a malicious computer program which runs in the background of the system. It may be a great

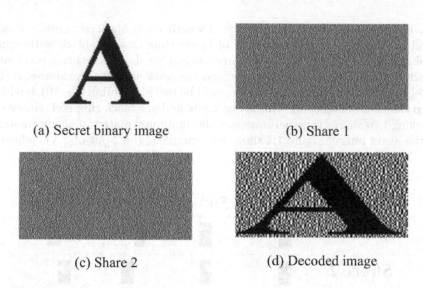

(a) Secret binary image (b) Share 1

(c) Share 2 (d) Decoded image

FIGURE 1.4 Example of 2-out-of-2 VC. Secret image is encoded into two random pattern and decoded image has 50 % contrast.

security risk for any individual or organization as the Trojan may send the confidential information from the host system to the owner of the Trojan's system. Let us consider a scenario where a user opens his bank account and gets a single physical transparency of a share which is generated by 2-out-of-2 VC scheme by bank administration. Another share will be kept by the bank itself in digital form. In order to secure online transactions, like online money transfers, the user gets a set of transparencies, each with a visual cryptography pattern printed on it, from the transaction server. Example 1.2 explains the modus operandi of visual cryptography in order to secure any transaction.

Example 1.2: A Trojan free secure authentication and transaction using VC are shown in Figure 1.5. Let us consider a scenario where a user opens his bank account and the bank authority generates two shares for a secret key using 2-out-of-2 VC. The secret key may be anything like an image, random patterns of shapes, different permutations of characters, etc. One physical transparency of the share will be given to the account holder whereas the other will be kept by the bank in digital form. Just like a one time password (OTP) whenever a user wants particular transaction, he needs to stack his own share with bank's share to reveal the secret key. Since in this application shares are in different mediums (one is in digital form and another is on hard transparency), the virtual stacking of shares are done for revealing the secure bank's instructions. This instruction may be like clicking on the given pattern or number by using a mouse. Here the Trojan on the computer is

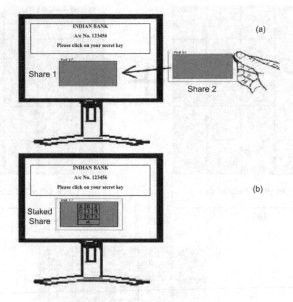

FIGURE 1.5 Securing the online transaction using VC: (a) To log in, the server sends an encrypted image of a permuted keyboard, which the user can only read after placing the slide over it, (b) The user enters the PIN by clicking at the positions according to their order in the PIN.

not able to see the permutation of characters chosen by the user which was randomly generated by the server because the mouse-clicks of the user cannot be interpreted by the Trojan. This method can be generalized to any alphabet and allows the sender to send short messages to the receiver in a secure way. It will be more secure when for each transaction we change the secret hence the share and its location on the computer screen. For this purpose we must have a share which allows decryption of more than one secret. Figure 1.6 shows the shares in which each grid contains a secret and that secret will only be revealed on its turn otherwise there will be random information. Here Figure 1.6 (a) represents share 1 which will be with the bank authority in digital mode and share 2 shown in Figure 1.6 (b) will be with the account holder in physical mode. When we stack both shares then a particular number of grids will prompt the secret as shown in Figure 1.6 (c) hence by this way the user can also track the number of transactions. Here the number of prompted grids will show the number of current transactions. Suppose when both shares are superimposed together and the secret is revealed on the third grid, it means two transactions have already been done. Figure 1.6 (c) shows that eighth number of transaction is going to be performed.

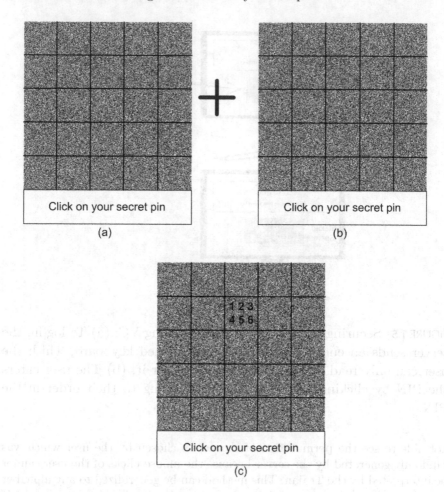

FIGURE 1.6 Shares with multiple secrets for secure bank transactions.

1.3.2 Authentication

Visual cryptography can be effectively used for authentication purposes. Here an administrator or trusted third party will generate two shares of the secret using 2-out-of-2 VC for each user. The administrator will keep one share along with the replica of the secret image and the user will be provided with a second share. At the time of authenticity verification, both shares will be stacked and compared with the stored secret image. An exact match between the recovered secret and the stored secret confirms the authenticity of the user.

1.3.3 Access Control

By using k-out-of-n visual cryptography, access for a particular system can be controlled. We can understand it by a simple analogy. Let us consider a joint bank account which is handled by n account holders, but we wish that any transaction will only be possible when for any fix value of k, $k > 1$ account holders will come together. If less than k account holders will come together then they will not be able to do any banking operations. This special kind of access control can easily be implemented by (k, n) VC.

1.3.4 Transaction Tracking

By VC we can track the leak point of any licensed software, made for the limited number of users, say n. n secret images are selected to protect the software. The ith copy of licensed software will be embedded with one of the shares of the ith secret image, generated by 2-out-of-2 VC. Other shares of all secret images along with the respective secrets will be stored with developer. If any pirated version of the software is found, then its extracted share will be stacked with each of the stored shares with the developer to find the culprit.

1.3.5 Watermarking

Using visual cryptography we can enhance the power and applicability of robust watermarking. Here instead of the ownership information (copyright logo) itself, share of ownership information is embedded into the cover image as a robust watermark. Let's say shares of copyright information are generated by 2-out-of-2 VC. One share will be embedded in the cover image using robust watermarking techniques and another will be kept by a trusted third party or owner of the cover image. Now we can securely do the ownership identification just by extracting the share from the watermarked image and stacking it with the owner's copy of the share. In spite of knowing the owner's copyright logo, the attacker cannot generate the share to create an objectionable image on behalf of the owner. The imperfect share generated by the attacker will never decode the exact copyright logo. One can understand this scenario through Figure 1.7, where (a) is the secret copyright logo and (b) and (c) are two shares of secret (a) generated by the 2-out-of-2 VC scheme. Now (d) is the watermarked image which is generated by the embedding of share 1 into the cover image. In a normal watermarking algorithm, we simply embed image (a) which can be reproduced by the attacker to misguide users. If we embed the share as a watermark, then it will be very tough to exactly regenerate the share by an attacker.

1.4 PRELIMINARIES

Let $W = \{W_0, ...W_{n-1}\}$ be a set of participants. A visual secret sharing scheme for a set W is a method to encode a secret binary image I_{sec} into n random

FIGURE 1.7 Watermarking using VC: (a) Copyright logo, (b) Share 1, (c) Share 2, (d) Watermarked image embedded with share 1.

images called shares, where each participant in W will receive one secret share. Let $\Gamma_{Qual} \subseteq 2^W$ and $\Gamma_{Forb} \subseteq 2^W$ where 2^W is power set of W and $\Gamma_{Qual} \cap \Gamma_{Forb} = \phi$. The members of Γ_{Qual} are referred to as the qualified set and members of Γ_{Forb} are referred to as the forbidden set. Γ_{Qual} is a set of all those subsets of shares by which the secret can be revealed. In (k, n) VC, any member of Γ_{Qual} is of size k or more. In other words, all subsets of shares of size greater or equal to k are members of Γ_{Qual}. Similarly Γ_{Forb} is a set of all those subsets of shares by which a secret cannot be revealed. In (k, n) VC, any member of Γ_{Forb} is of size less than k. In other words, all subsets of shares of size less than k are members of Γ_{Forb}. The pair $(\Gamma_{Qual}, \Gamma_{Forb})$ is called the access structure of the VC. If the participants in a subset $X \in \Gamma_{Qual}$, the participants in any superset Y of $X(X \subset Y)$ should be able to decode the secret image as well, hence $Y \in \Gamma_{Qual}$. Therefore Γ_{Qual} is called monotone increasing. Similarly, if subset $X \in \Gamma_{Forb}$, the participants in any set Y of

$X(Y \subset X)$ should not be able to decode the secret image either. Hence, Γ_{Forb} is called monotone decreasing.

Let $\Gamma_0 = \{X | X \in \Gamma_{Qual} \ \& \ \forall Y \subset X \Rightarrow Y \notin \Gamma_{Qual}\}$ be the set of all minimally qualified subsets. In a strong access structure, Γ_{Qual} is the closure of Γ_0 . Thus, Γ_0 is termed a basis, from which a strong access structure can be derived. Any qualified set of participants $X \in \Gamma_{Qual}$ can visually reveal the I_{sec} but any participants $Y \in \Gamma_{Forb}$ cannot.

Example 1.3: The strong access structure $(\Gamma_{Qual}, \Gamma_{Forb})$ of the $(2,3)$ VC scheme can be obtained as $\Gamma_{Qual} = \{\{1,2\}\{2,3\}\{1,3\}\{1,2,3\}\}$ and $\Gamma_{Forb} = \{\phi, \{1\}, \{2\}, \{3\}\}$. It can be verified that Γ_{Qual} is monotone increasing and Γ_{Forb} monotone decreasing. Let $X = \{1,2,3\} \in \Gamma_{Qual}$ and $Y = \{1,2\} \subset X$. Since $Y \in \Gamma_{Qual}$, it follows that X is not in Γ_0, that is, $X \notin \Gamma_0$. Now let $X = \{1,2\} \in \Gamma_{Qual}$. Any $Y \subset X$ satisfies $Y \notin \Gamma_{Qual}$, so $X \in \Gamma_0$. The same result can be obtained on $\{2,3\}$ and $\{1,3\}$. Hence, $\Gamma_0 = \{\{1,2\}\{2,3\}\{1,3\}\}$.

Example 1.4: For the 2-out-of-2 scheme shown in Figure 1.4, the set of participants W will be $\{W_0, W_1\}$ because there are only two participants. Hence, $\Gamma_{Qual} = \{\{W_0, W_1\}\}$ and $\Gamma_{Forb} = \{\{W_0\}\{W_1\}\}$.

1.5 FUNDAMENTAL PRINCIPLES OF VISUAL SECRET SHARING

When we develop a visual cryptography scheme, we need to characterize it by three parameters as shown in Figure 1.8. All of these parameters have relationships with each other, for example, there is a trade-off between pixel expansion and contrast.

1.5.1 Pixels Expansion m

Pixel expansion is the number of subpixels in which single pixel p of the secret image I_{sec} is encoded in every share.

Example 1.5: For the 2-out-of-2 scheme shown in Figure 1.4, the pixel expansion m is 2 because each pixel, either 1 or 0 of the secret, is encoded by two subpixels in each share. That is why in this case the dimension of shares and the secret will mismatch and the shares will be larger than the secret which is not desirable.

Pixel expansion must be kept to a minimum as much. The ideal value for pixel expansion is $m = 1$. In this case one pixel of the secret image will be

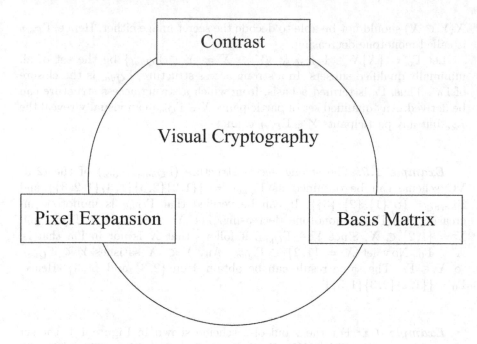

FIGURE 1.8 Characteristics of visual cryptography.

encoded by one pixel in each share, hence the dimension of secret image and its shares will be the same.

1.5.2 Contrast α

As in the traditional VC, black pixels are recovered with 100% accuracy after stacking the qualified number of shares while white pixels are not, hence contrast is the measurement of the visibility for recovered white pixels in the decoded image. Contrast is a very important measure to judge the effectiveness of any VC approach because there is a trade-off between security and contrast. To achieve the security, all state of art approaches on VC suffer from some contrast loss. There is also a trade-off between pixel expansion and contrast. When we increase the pixel expansion, then the value of the contrast will decrease. The contrast value for any VC approach must be as high as possible.

Example 1.6: For the 2-out-of-2 VC scheme shown in Figure 1.4, black pixels are recovered with 100% accuracy after stacking both shares, but white pixels are recovered with only 50% confirmation. This scenario is called contrast loss. In this example, white pixels are recovered in a combination of one black and one white subpixel. So 50% of the white pixels of the original secret continue to be white in the recovered secret. Hence, the visibility of white pixels in the decoded image will be, 1/2 i.e., 50% contrast.

1.5.3 Basis Matrices

Basis matrices are used to construct the shares. Two matrices S^0 and S^1 of size $n \times m$ (where n and m are the number of shares and pixel expansion, respectively) are called basis matrices for bit 0 and 1, respectively, if two collections C_0 and C_1 containing matrices obtained by permuting in all possible ways the columns of S^0 and S^1, respectively, satisfy some conditions (explained in the next section). As all traditional VC methods are based on a binary secret image, that is why only two basis matrices are generated, one is for black and the other one is for the white pixel.

Example 1.7: Figure 1.9 shows the example of a basis matrix for the given 2-out-of-2 VC scheme shown in Figure 1.4. For coding 0, any one matrix of C_0 is selected and any one row of it is assigned in the first share and the other row to the second share. Similarly for coding 1, one matrix of C_1 is selected and any one row of it is assigned to the first share and the remaining row to the other share.

$$S^0 = \begin{bmatrix} 0 & 1 \\ 0 & 1 \end{bmatrix}, \quad S^1 = \begin{bmatrix} 0 & 1 \\ 1 & 0 \end{bmatrix}$$

$$C_0 = \left\{ \begin{bmatrix} 0 & 1 \\ 0 & 1 \end{bmatrix}, \begin{bmatrix} 1 & 0 \\ 1 & 0 \end{bmatrix} \right\}$$

$$C_1 = \left\{ \begin{bmatrix} 0 & 1 \\ 1 & 0 \end{bmatrix}, \begin{bmatrix} 1 & 0 \\ 0 & 1 \end{bmatrix} \right\}.$$

FIGURE 1.9 Example of a basis matrix for 2-out-of-2 VC.

Points worth remembering:
There are three main parameters, namely, basis matrix, pixel expansion and contrast which must be characterized for any VC approach.

Points worth remembering:
There is a trade-off between contrast and pixel expansion.

1.5.4 Concept of Black and White Pixels in Visual Cryptography

Basically, in image processing, for binary images, the intensity value 1 is treated as white and the intensity value 0 is treated as black. This concept is not true in visual cryptography. In VC, a white pixel in denoted by 0 and a black pixel is denoted by 1. There is a valid reason for this reverse concept in visual cryptography. Share staking is nothing but the resemblance of a logical OR operation. According to the truth table of a logical OR,

$$
\begin{vmatrix}
1 & OR & 1 & \rightarrow & 1 \\
1 & OR & 0 & \rightarrow & 1 \\
0 & OR & 1 & \rightarrow & 1 \\
0 & OR & 0 & \rightarrow & 0
\end{vmatrix}
$$

Here we can see that the binary value 1 always dominates in the result of the OR operation. When a share is printed on a transparency, then the black intensity of the share is printed by black ink whereas the white intensity of the share is left blank (transparent). Whenever we superimpose two or more shares and if there is any black intensity at any location of the some shares, then it will dominate the other transparent portions of the shares which denote the white intensities. It means while stacking, black intensity plays the role as 1 in logical OR operations. That is why in VC, black pixels are denoted by 1, whereas white pixels are denoted by 0.

Points worth remembering:
In VC, if binary images are used as the secret then 1 denotes black and 0 denotes the white pixel.

Points worth remembering:
Staking of the shares resembles a logical OR operation.

1.6 FORMATION OF A BASIS MATRIX

Shamir has proposed a very efficient scheme to generate a basis matrix for the (n, n) visual secret sharing scheme. Let a (n, n) VC scheme and W be the ground set given by $W = \{W_1, W_2..W_n\}$ of n participants. Let $\pi = \{\pi_1, \pi_2..\pi_{2^{n-1}}\}$ be the set of all subsets of even cardinality and $\sigma = \{\sigma_1, \sigma_2..\sigma_{2^{n-1}}\}$ be the set of all subsets of odd cardinality. Basis matrix S^0 and S^1 are constructed by π and σ, respectively. All matrices are obtained by permuting the column of $S^0[S_{ij}^0]$ and $S^1[S_{ij}^1]$ for $1 \leq i \leq n$ and $1 \leq j \leq 2^{n-1}$, in which $S_{ij}^0 = 1$ if and only if $w_i \in \pi_j$ and $S_{ij}^1 = 1$ if and only if $w_i \in \sigma_j$. Hence, the size of basis matrix S^0 and S^1 will be $n \times m$ where pixel expansion $m = 2^{n-1}$.

Example 1.8: Construction of a basis matrix for a $(4,4)$ VCS by Noar and Shamir.

Here $n = 4$, and the set of participants $W = \{W_1, W_2, W_3, W_4\}$

$$\pi = \{\phi\}\{W_1W_2\}\{W_2W_3\}\{W_3W_4\}\{W_1W_3\}\{W_1W_4\}\{W_2W_4\}\{W_1W_2W_3W_4\}$$
$$\sigma = \{W_1\}\{W_2\}\{W_3\}\{W_4\}\{W_1W_2W_3\}\{W_2W_3W_4\}\{W_1W_3W_4\}\{W_1W_2W_4\}$$

Now according to Noar and Shamir's scheme the size of the basis matrix will be $n \times 2^{n-1}$, i.e, 4×8 where $m = 2^{n-1}$ for the given example.

$$S_{ij}^0 = 1, \text{ If } W_i \in \pi_j$$
$$\text{and}$$
$$S_{ij}^1 = 1, \text{ If } W_i \in \sigma_j$$

where $1 \leq i \leq n$ and $1 \leq j \leq 2^{n-1}$. π and σ are used to construct the basis matrix S^0 and S^1, respectively.

$$S^0 = \begin{bmatrix} 01001101 \\ 01100011 \\ 00111001 \\ 00010111 \end{bmatrix} \quad S^1 = \begin{bmatrix} 10001011 \\ 01001101 \\ 00101110 \\ 00010111 \end{bmatrix}$$

For construction of S^0, let $i = 2$ then $W_2 \in \pi_2, \pi_3, \pi_7, \pi_8$ in set π, therefore $S_2^0 = 01100011$. Similarly for construction of S^1, let $i = 2$ then $W_2 \in \sigma_2, \sigma_5, \sigma_6, \sigma_8$ in set σ, therefore $S_2^1 = 01001101$. In this example pixel expansion $m = 2^{n-1} = 8$. It means, the single pixel is denoted by eight sub-pixels and contrast $\alpha = \frac{1}{2^{n-1}} = \frac{1}{8}$ which is too low. This situation will be worse when the number of participants n is larger.

1.6.1 Observations Related to The Basis Matrix Creation Approach of Naor and Shamir

The following points are observed during creation of basis matrices using Naor and Shamir's approach for n number of shares.

1. Each row of both S^0 or S^1 contains n 1s and n 0s.

2. OR operation on rows of S^0 contain single 0 in the resultant vector.

3. OR operation on rows of S^1 contain all 1s in the resultant vector.

4. Pixel expansion is directly dependent on the number of participants as $m = 2^{n-1}$, so if n is large then pixel expansion will be exponentially large. In given example one pixel is denoted by 2^{n-1} subpixels.

5. Contrast α also depends on the number of participants as $\alpha = \frac{1}{2^{n-1}}$. If the value of n increases α decreases. In the given example, the value of α is $\frac{1}{8}$ which is very low.

Now basis matrices are used to construct the shares. The first two sets C_0 and C_1 are obtained which have matrices by permuting in all possible ways the columns of S^0 and S^1, respectively.

Now each row of the basis matrix is used to construct the individual share. For example, row one will be assigned to the first participant, row two will be assigned to the second participant and so on. A basis matrix will be made for each pixel of the secret image.

1.6.2 Essential Conditions for a Basis Matrix

Basis matrices for the intensities 0 and 1 are basic units for any visual cryptography scheme. Shamir has suggested some validity testing for any pair of matrices to be basis matrices. In VC each secret binary pixel p is converted in to m subpixels for each of the n shares. A Boolean matrix M of size $n \times m$ is used to describe these subpixels where value 0 is used to denote a white subpixel and value 1 is used for a black subpixel. Let r_i be the $i^{th} (i = 1, 2, 3.., n)$ row of M which contains subpixels for the ith share. The hamming weight $w(v)$ of binary vector v is proportional to the visual intensity of p where $v = OR(r_{i_1}, r_{i_2}..r_{i_s})$ (bitwise logical OR) and $r_{i_1}, r_{i_2}...r_{i_s}$ are the rows of M.

Definition-1.1: Let $(\Gamma_{Qual}, \Gamma_{Forb})$ be an access structure for n participants. The collections of boolean matrices C_0 and C_1, each of dimension $n \times m$ constitute a VC scheme where m denotes pixel expansion, if there exist $\alpha(m)$ and t_X for every $X \in \Gamma_{Qual}$ satisfying the following two conditions:

1. **Contrast Condition:** Any qualified subset $X = \{i_1, i_2..., i_u\} \in \Gamma_{Qual}$ can decode the I_{sec} by stacking the respective transparencies. Formally, for matrix $M \in C_j, (j = 0, 1)$, the row vector $v_j(X, M) = OR(r_{i_1}, r_{i_2}..r_{i_s})$. It holds that

$$w(v_0(X, M)) \leq t_X - \alpha(m) \times m \quad \forall m \in C_0 \qquad (1.1)$$

and

$$w(v_1(X, M)) \geq t_X \quad \forall m \in C_1 \qquad (1.2)$$

where t_X is the threshold to visually interpret the reconstructed pixel as black or white and $\alpha(m)$ is called the relative difference referred to as the contrast of the decoded image. These can be obtained by

$$t_X = min(w(V_1(X, M))) \qquad (1.3)$$

where $M \in C_1$

$$\alpha(m) = \frac{min(w(V_1(X, M))) - max(w(V_0(X, M)))}{m} \qquad (1.4)$$

Example 1.9: In the continuation of example 1.8, the contrast condition can be verified for basis matrix S^0 and S^1. $t_X = min(w(V_1(X, M)))$ where $M \in C_1$, since here we are taking all rows of M so $\forall w(V_1) = \{8\}$ hence $t_X = 8$. Now $max(w(V_0)) = 4$. Since pseudo pixel expansion $m = 8$ hence $\alpha(m) = \frac{1}{2}$. For the contrast condition, $w(V_0) \leq t_X - \alpha(m).m$ and $w(V_1) \geq t_X$ must be satisfied, that is, $4 \leq 8 - \frac{1}{2}.8$ or $4 \leq 4$ and $8 \geq 8$.

2. **Security Condition:** Any forbidden subset $X = \{i_1, i_2 ..., i_v\} \in \Gamma_{Forb}$ of v participants has no information related to the secret image.

Example 1.10: In the continuation of example 1.8, the security condition can be verified for basis matrix S^0 and S^1. The given approach is a 4-out-of-4 scheme, hence the set of any three shares belongs to the forbidden set. Any three rows from S^0 or S^1 will provide redundant bits and hence that will be indistinguishable. If we take the bitwise OR of the first three rows of both matrices S^0 and S^1 we will get different bit permutations of bit stream 01111111. Hence, it will be very difficult to identify the corresponding secret bit.

The basis matrix for any access structure $(\Gamma_{Qual}, \Gamma_{Forb})$ must satisfy both the above mentioned conditions for any visual secret sharing scheme.

Points worth remembering:
A basis matrix will only be valid if it satisfies the contrast and security conditions.

1.7 DIFFERENT EVALUATION PARAMETERS

The efficiency and accuracy of a visual cryptography approach can be evaluated by various parameters. These evaluation parameters are classified into two categories as shown in Figure 1.10.

1.7.1 Objective Evaluation Parameters

Traditional and most of the state-of-the-art VC approaches are based on the protection of the binary secret images. Objective evaluation parameters are

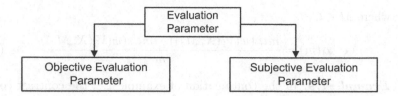

FIGURE 1.10 Categorization of evaluation parameters.

perfectly suitable for binary images because they depend on one-to-one correspondence between pixels, their values and locations. There are various error parameter measurements used for quantitative evaluation.

Let us suppose that I and G are input and output binary images, respectively. N_{TP}, N_{FP}, N_{TN} and N_{FN} are represented as true positive, false positive, true negative, false negative with respect to I and G, respectively. Here N_{TP}, N_{FP}, N_{TN} and N_{FN} are defined as:

True Positive (N_{TP}): When the binary pixel value 1 of the input image remains the same as 1 into the output image, this case is called true positive and the number of such type of pixels is denoted by N_{TP}.

False Positive (N_{FP}): When the binary pixel value 0 of the input image is altered as 1 into the output image, that case is called false positive and the number of such type of pixels is denoted by N_{FP}.

True Negative (N_{TN}): When the binary pixel value 0 of the input image remains the same as 0 into the output image that case is called true negative and the number of such type of pixels is denoted by N_{TN}.

False Negative (N_{FN}): When the binary pixel value 1 of the input image is altered as 0 into the output image that case is called false negative and the number of such type of pixels is denoted by N_{FN}.

Points worth remembering:
All objective evaluation parameters are calculated through four parameters viz false positive, false negative, true positive and false negative.

On the basis of this set of four numbers, N_{TP}, N_{FP}, N_{TN} and N_{FN}, there are many error metrics calculated for evaluation of similarity between two binary images. Some of the most important parameters are discussed below:

Negative Rate Matrix (NRM): The NRM depends on the pixel wise inequality between the I and G.

$$NRM = \frac{NR_{fn} + NR_{fp}}{2} \tag{1.5}$$

where

$$NR_{fn} = \frac{N_{FN}}{N_{FN} + N_{TP}} \tag{1.6}$$

$$NR_{fp} = \frac{N_{FP}}{N_{FP} + N_{TN}} \tag{1.7}$$

The value of NRM between two similar images is 0.

Recall/Sensitivity:

$$Recall = \frac{N_{TP}}{N_{TP} + N_{FN}} \tag{1.8}$$

The value of recall between two similar images will be 1.

Precision:

$$Precision = \frac{N_{TP}}{N_{TP} + N_{FP}} \tag{1.9}$$

The value of precision between two similar images will be 1.

F-Measure:

$$FM = \frac{2 \times Recall \times Precision}{Recall + Precision} \tag{1.10}$$

The value of F-Measure between two similar images will be 1.

Specificity:

$$Specificity = \frac{N_{TN}}{N_{TN} + N_{FP}} \tag{1.11}$$

The value of specificity between two similar images set to be 1.

Balanced Classification Rate (BCR)/Area Under the Curve (AUC):

$$BCR = 0.5 \times (Specificity + Sensitivity) \tag{1.12}$$

The value of BCR/AUC between two similar images will be 1.

Balanced Error Rate (BER):

$$BER = 100 \times (1 - BCR) \tag{1.13}$$

The value of the balanced error rate (BER) between two similar images is set to be 0.

Structural Similarity Index (SSIM):
SSIM measures are based on the Human Visual System (HVS). For measuring the image quality different types of HVS models have been proposed. However, the SSIM measure is an objective parameter which is similar to the subjective evaluation parameter.

$$SSIM(x, y) = \frac{(2\mu_x\mu_y + c_1)(2\sigma_{xy} + c_2)}{(\mu_x^2 + \mu_y^2 + c_1)(\sigma_x^2 + \sigma_y^2 + c_1)} \tag{1.14}$$

where μ_x, μ_y, σ_x^2, σ_y^2 and σ_{xy} refer to the value of average, variance and covariance for x and y, respectively. The value of SSIM varies between -1 and 1, where the maximum value, i.e., 1 is obtained for two similar images.

Distance-Reciprocal Distortion Measure (DRDM):
Let W_m be the weight matrix and i_c and j_c are the center pixel.

$$W_m(i, j) = \begin{cases} 0, & \text{if } i_c = j_c \\ \frac{1}{\sqrt{(i-i_c)^2+(j-j_c)^2}}, & \text{otherwise} \end{cases} \tag{1.15}$$

This matrix is normalized by.

$$W_{Nm}(i, j) = \frac{W_m(i, j)}{\sum_{i=1}^{m} \sum_{j=1}^{m} W_m(i, j)} \tag{1.16}$$

Now

$$DRD_k = \sum_{i,j} [D_k(i, j) \times W_{Nm}(i, j)] \tag{1.17}$$

where D_k is given by $(B_k(i, j) - g[(x, y)_k])$. Thus DRD_k equals the weighted sum of the pixels in the block B_k of the original image.

$$DRD = \frac{\sum_{k=1}^{s} DRD_k}{NUBN} \tag{1.18}$$

where NUBN are the nonuniform blocks in $F(x, y)$. For an identical image DRDM will be 0.

A visual cryptography algorithm may be effective, if the values of objective evaluation parameters reach towards their ideal values, during the comparison between recovered secret and original secret.

1.7.2 Subjective Parameters

Besides the aforementioned objective evaluation parameters used for finding the best VC algorithm, there are some other essential characteristics of visual cryptography called subjective parameters, which can also be utilized for comparison of VC algorithms. One can design a VC algorithm for achieving the ideal value for any one of the following characteristics. These are nothing but research issues because all existing state of art approaches on VC are working toward their improvement.

1. **Pixel Expansion** m: Pixel expansion for secret pixel p denotes the number of encoding subpixels by which p is encoded. To preserve the dimension of shares as of secret image I_{sec}, we need to choose the value of m that is as small as possible.

2. **Decoding Process:** As per the properties of visual cryptography, the secret image should be obtained without any cryptographic computation. So in this case only the HVS is sufficiently capable to decode the secret. But in some cases, where a gray scale secret image is taken as an input, a little bit of computation is required at the receiver end in order to decode the secret.

3. **Contents of Shares:** A meaningless share is a common problem in most of the existing visual cryptographic algorithms. These types of shares are more vulnerable to cryptographic attacks and cryptanalysis. They also create problems in share identification if a large number of shares is there. That is why some meaningful information related to the share holder is desired in the share.

4. **Contrast** $\alpha(m)$ **of the Decoded Image:** In order to create secure shares, most of the VC approaches suffer from the contrast loss problems. A good VC approach ensures a high value for contrast $\alpha(m)$ so that the visual quality of the secret image and its decoded version remain same.

5. **Security Criteria:** Γ_{Forb} or its subset must be clueless about the secret image. Γ_{Forb}, rows from any basis matrix $C_{j(j\in\{0,1\})}$ must be indistinguishable with respect to p.

6. **Codebook Requirement:** Many VC algorithms require a codebook in order to encode and decode the secret. A codebook is a collection of all combinations of C_0 and C_1 for different variations of pixels. Unfortunately, a large codebook is very difficult to handle and uses a large static memory. Hence, we need some algorithms which generate a codebook implicitly.

7. **Limit of Shares:** The number of shares must be variable in nature so that The VC algorithms are generalized enough to be used for any kind of security application.

8. **Security of Shares:** Shares carry secret information, hence, they must be verifiable in nature. A verifiable share is capable enough to authenticate itself at the time of any contradiction. There are various ways to make shares verifiable. Self-embedding is an efficient way to obtain the objective of verifiability as no extra image is required in this case.

9. **Copyright Information:** A share may be embedded with a hidden copyright related to the credentials of the share holder and it can be extracted at the receiver end for ownership assertion.

10. **Secret Payload:** In many cases additional abstract information related to decryption keys or a technique at the receiver end may be required. Hence, shares must have some additional payload capacity in which all these information can be embedded.

11. **Multiple Secrets:** Though n shares perfectly encrypt and decrypt one secret, they consume more bandwidth and take more transmission time when secretly transferring secret data.

12. **Type of Secret Image:** Most of the state of art approaches deal with binary images as secret but binary images are inadequate for the number of security applications. Hence, the VC algorithm must be extendable for multi-tone images also.

13. **Progressive Recovery:** Sometimes availability of all k participants in k-out-of-n visual cryptography is not possible (if k is large) and hence nothing will be revealed. To avoid this situation the VC must be designed in such a way that secret can also be recovered progressively with fewer number of shares.

There are many existing state of art approaches in VC which have tried to achieve subjective evaluation parameters ideally as much as possible and they have succeeded to some extent. Some subjective evaluation parameters are conflicting in nature such as when the secret is a multi-tone image then we cannot decode it just by the stacking of shares. To decode the secret at the receiver end, a little bit of computation is required. Hence, due to the distinctive nature of each subjective evaluation parameter, visual cryptography is further classified in various dimensions which we will see in the next chapter.

Points worth remembering:
Visual cryptography is classified on the basis of various subjective evaluation parameters.

SUMMARY

- A *secret* may be in the form of a text, numeric or multimedia file.
- *Visual cryptography* is a cryptographic scheme used to protect or secretly share the images.
- In traditional visual cryptography, secret decryption is done with *no computation.*
- Visual cryptography can be used for enhancing the scope of security for many practical applications in real life like watermarking, Trojan free secure transactions, authentication, access control etc.
- There are three main parameters viz *basis matrix, pixel expansion and contrast* which must be characterized for any VC approach.
- There is a trade-off between contrast and pixel expansion. .
- In VC, if binary images are used as the secret, then 1 denotes the black and 0 denotes the white pixel.
- Staking of the shares resembles the *logical OR operation.*
- A *basis matrix* will only be valid if it satisfies the contrast and security conditions..
- All objective evaluation parameters are calculated through four parameters viz *false positive, false negative, true positive and false negative.*
- Visual cryptography is classified on the basis of various subjective evaluation parameters.
- *Objective measures* are well suited for binary images.

CHAPTER 2

Various Dimensions of Visual Cryptography

CONTENTS

I N order to achieve the ideal values of subjective evaluation parameters, visual cryptography is categorized into various subcategories. There are many subjective evaluation parameters which are conflicting in nature and cannot be achieved in a single VC approach simultaneously. For example in the previous chapter we saw that a good VC approach must have computationless recovery (just by stacking of shares) of the secret image which is only possible when we deal with the binary secret images. We also saw that binary images are inadequate for the number of security applications. Hence, the VC algorithm must be extendible for multi-tone/color images also. In the case of color images, we cannot recover the secret just by stacking the shares. In this case, we need a little bit of computation at the receiver end. This conflict in the requirement is responsible for the division of VC approaches into two categories which are VC for binary images and VC for color images. This is not the only reason for categorization of VC approaches. There are many others which we will see in this chapter.

The **Chapter Learning Outcomes (CLO)** for this chapter are given below. After reading this chapter, readers will be able to:

1. Understand the various categories of visual cryptography.

2. Understand the need, advantages and disadvantages of each category of visual cryptography.

3. Identify the research gaps in various dimensions of visual cryptography.

2.1 VARIOUS DIMENSIONS OF VISUAL CRYPTOGRAPHY

The classification of various dimensions of visual cryptography is shown in Figure 2.1. We can see here that visual cryptography can be broadly classified into three categories viz. Traditional Visual Cryptography (TVC), Extended Visual Cryptography (EVC) and Dynamic Visual Cryptography (DVC). Each category has its own subcategories which may be called dimensions of visual cryptography. Each dimension has some unique characteristics that are specifically used for particular applications. These classifications of the dimensions

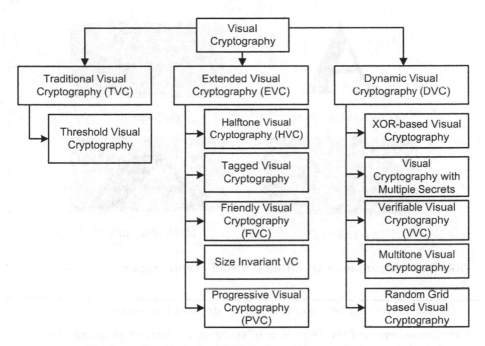

FIGURE 2.1 Various dimensions of visual cryptography.

may be on the basis of achieving the ideal values of one or more subjective evaluation parameters for a particular application.

Points worth remembering:
Visual Cryptography schemes are broadly classified into three categories viz. Traditional Visual Cryptography (TVC), Extended Visual Cryptography (EVC) and Dynamic Visual Cryptography (DVC).

2.1.1 Traditional Visual Cryptography (TVC)

As we know, Naor and Shamir first invented the concept of visual cryptography. All those approaches which follow the basic principle of Shamir's approach are called Traditional Visual Cryptography (TVC) approaches. When we say that an approach is traditional it means that it is following the basis matrix creation approach of Shamir. All shares which are obtained by Shamir's basis matrices have the following properties:

1. The dimensions of all shares will be larger than the original secret image.

2. Each share will be visually random in nature.

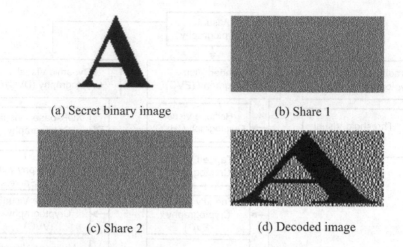

(a) Secret binary image
(b) Share 1

(c) Share 2
(d) Decoded image

FIGURE 2.2 Example of traditional visual cryptography.

3. The secret can be recovered just by stacking the shares.

One can understand the behaviour of traditional visual cryptography by Figure 2.2. Here we can see that the dimensions of the shares are $m \times 2n$ for the secret of dimension $m \times n$. Similarly both shares are visually random in nature. Actually in this figure, both shares are generated by the basis matrices from Shamir.

Points worth remembering:
All those approaches come under the category of traditional VC which follows the basis matrices of Shamir's approach.

2.1.1.1 Threshold visual cryptography

Generalization of traditional visual cryptography is called threshold visual cryptography. A threshold VC scheme is denoted as (k, n)-VCS, where k is the threshold value and n is the number of participants. The significance of threshold k in threshold visual cryptography is that if less than k number of participants meet together then they cannot decode the secret in spite of having infinite computational power. The secret will only be revealed when the number of participants is k or more. That is why the threshold k-out-of-n visual cryptography (VC) approach is known as an "All or Nothing" scheme. $2 \leq k \leq n$, is a very important relation between k and n. Figure 2.2 is a very trivial example of threshold visual cryptography where $k = n = 2$. Most of the early research on VC starting from Shamir's approach is considered to

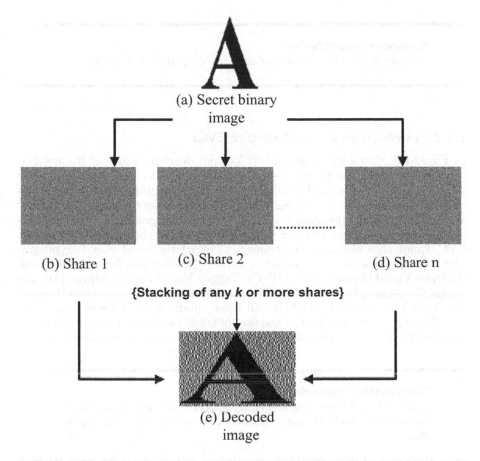

FIGURE 2.3 Example of threshold visual cryptography.

be threshold visual cryptography. After Shamir's contribution in the field of VC, most of the researchers enhanced the power of the basis matrix in order to get a generalized scheme for k-out-of-n or (k, n) VC scheme. Figure 2.3 demonstrates the concept of a (k, n) VC scheme. Here we can see that the secret will only be revealed when k or more than k number of shares are stacked together.

All available threshold visual cryptography schemes are based on general access structures. Here, k is responsible for making the forbidden and qualified subset of shares. The number of elements (shares) in the forbidden subset will always be less than k whereas in the qualified subset it will greater than or equal to k. All the forbidden and qualified subsets of these shares are a predefined set of the access structure. The participants maintained in a forbidden subset are not allowed to recover the secret while participants listed in a qualified subset have authority for the same.

Points worth remembering:
A threshold VC scheme is defined as (k, n)-VCS, where $2 \leq k \leq n$.

2.1.2 Extended Visual Cryptography (EVC)

In Extended Visual Cryptography (EVC) approaches, many additional features are added to traditional visual cryptography. EVC approaches preserve the very basic principles of traditional visual cryptography but improves its applicability by extending its features. For example we can say that all EVC approaches preserve the main feature of visual cryptography that is computationless recovery at the receiver end just by stacking the shares. Most of the EVC approaches also hold the concept of (k, n) threshold visual cryptography. Extended visual cryptography is further classified into five categories viz. Halftone Visual Cryptography (HVC), Tagged Visual Cryptography, Friendly Visual Cryptography (FVC), Size Invariant Visual Cryptography and Progressive Visual Cryptography (PVC). All these subdivisions of visual cryptography slightly modify/extend the features of traditional visual cryptography by maintaining its basic principles, which we will see in the next subsection.

Points worth remembering:
Extended Visual Cryptography (EVC) approaches slightly modify/extend the features of traditional visual cryptography by maintaining its basic principles.

2.1.2.1 Halftone visual cryptography (HVC)

First of all, we must know what is a halftone image and how it is generated? Traditional visual cryptography schemes were based on binary secret images. Binary and halftone images both consist of only two intensities, one is for black and other is for white. But the appearance of each image is entirely different. A binary image may contain continuous clusters of black and white pixels withot having shades. Whereas halftone images are generated by the reprographic technique that simulates continuous/multitone images through the use of dots, varying either in size or in spacing, thus the image looks like a gray scale image. We can understand it by Figure 2.4, here (a) is the gray scale image of Lena which consists of 256 gray scale intensities. Figure 2.4 (b) is the binary version of the Figure 2.4 (a) which is achieved by the thresholding method. Figure 2.4 (c) is the halftone version of Figure 2.4 (a). We can see clearly that (c) is also made up of dots of only two intensities (black and white) with varying shapes and spacing. A halftone image creates an illusion of a gray scale image, hence it is useful in various applications.

Points worth remembering:
Halftone images are binary images with the illusion of a gray scale image.

2.1.2.2 Significance of a halftone image over a binary image

As we know, both halftone and binary images consist of only two intensity values but they look entirely different. Figure 2.4 (b) is an example of a binary image which is obtained by the thresholding method. Here we can see that the image contains cluster of either complete black or complete white intensities. There is no randomness in the distribution of the black and white pixels, hence this pattern is more predictable for cryptanalysis. If a pixel is black/white then there is a high probability of the neighboring pixels being black/white. Figure 2.4 (c) is an example of a halftone image which has more randomness then normal binary image. Prediction of pixel intensity is tough in this case because if a pixel is black/white then there is equal probability of neighboring pixels being black/white. If prediction of pixel intensity is tough, then the prediction of intensity by the pixel's encoded version is even tough. That is why halftone images for secrets are more suitable for visual cryptographic applications.

Points worth remembering:
Halftone images are more appropriate for visual cryptography in comparison to thresholded binary images because of their randomness.

2.1.2.3 Halftone image creation using error diffusion

Many methods exist in the literature to convert a multitone/continuous tone image into a halftone image. Error diffusion is one of the best and efficient algorithms to do this. This algorithm is based on a feedback approach where the quantization error at each pixel of an image is filtered and fed back to a set of future input intensity samples.

Figure 2.5 shows a binary error diffusion diagram where $f(m,n)$ shows the $(m,n)^{th}$ pixel of the input multitone image, $d(m,n)$ is the sum of the diffused past errors and the input pixel value $f(m,n)$, whereas $g(m,n)$ is the output halftoned pixel value. Error diffusion consists of mainly two components. The first component is the thresholding block where the output $g(m,n)$ is decided

FIGURE 2.4 Example of binary and halftone image: (a) Original gray scale image, (b) Binary image, (c) Halftone image.

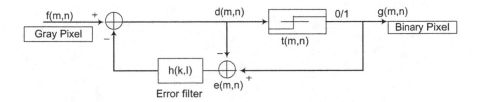

FIGURE 2.5 Block diagram for binary error diffusion: The pixel $f(m,n)$ is passed through a quantizer to get the corresponding pixel of the halftone $g(m,n)$. The difference between these two pixels is diffused to the neighboring pixels by means of the filter $h(k,l)$.

by

$$g(m,n) = \begin{cases} 1, & \text{if } d(m,n) \geq t(m,n) \\ 0, & \text{otherwise} \end{cases} \tag{2.1}$$

The threshold $t(m,n)$ can be position-dependent. The second component is the error filter $h(k,l)$ whose input $e(m,n)$ is the difference between $d(m,n)$ and $g(m,n)$. Finally, we can compute $d(m,n)$ as

$$d(m,n) = f(m,n) - \sum_{k,l} h(k,l)e(m-k,n-l) \tag{2.2}$$

The Floyd-Steinberg error filter is widely used for error diffusion. The weights of the filter are given by $h(0,1) = 7/16$, $h(1,-1) = 3/16$, $h(1,0) = 5/16$ and $h(1,1) = 1/16$. The recursive structure of the block diagram shown in Figure 2.5 indicates that the quantization error $e(m,n)$ depends not only on the current input and output but also on the entire past history. The error filter is designed in such a way that the low frequency difference between the input and output image is minimized.

Figure 2.6 shows the example of halftone visual cryptography where the secret image is first of all converted into its halftone version. Halftone images will be more suitable when we deal with natural images as the secret for visual cryptography. Figure 2.6 (a) shows the halftone version of Lena which is treated as the secret. (b) and (c) are the shares which are random as well as larger with respect to the secret in terms of dimension. The recovered image is denoted by (d) which is darker than image (a) because of contrast loss.

After the introduction of the concept of halftoning, whenever we deal with a natural image as the secret in visual cryptography, first of all the image is converted into its halftone version. That is why, at this point the default format for secret image will be a halftone image until noted explicitly for other formats.

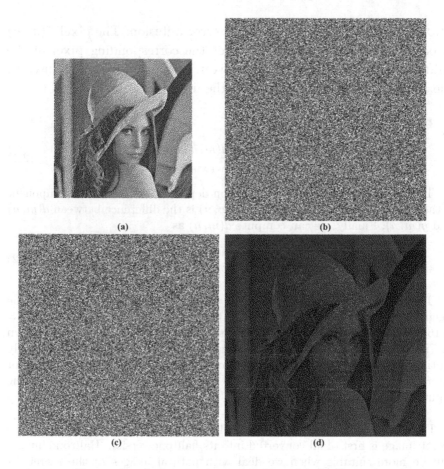

FIGURE 2.6 Example of visual cryptography with halftone image as secret: (a) Secret halftone image, (b) Random Share 1, (c) Random Share 2, (d) Recovered secret by stacking of share 1 and share 2.

Points worth remembering:
Error diffusion is a well-known method of halftoning which uses the Floyd-Steinberg error filter.

2.1.2.4 Tagged visual cryptography (TVC)

In a (k, n) visual cryptography scheme, if the values of k and n are too large then it will be very difficult to manage the random shares. Random shares will create confusion in share identification because each will look like a noisy image. Tagged Visual Cryptography (TVC) is a method which addresses this problem by providing extra identification information as a tag on the shares. All basic characteristics of traditional visual cryptography are preserved in addition to this tagged information.

One can understand the concept of Tagged Visual Cryptography (TVC) by Figure 2.7. Figure 2.7(a) shows the secret binary image and (b), (c) show the tag images for share 1 and share 2 where share 1 and 2 are represented by Figure 2.7 (d) and (e,) respectively. When we directly stack both the shares then we will get the recovered secret image shown in Figure 2.7 (f). In order to find out the identification of the shares we fold the shares from mid-point. As we can see in Figure 2.7 (g) and (h) after folding the shares we get the tagged image. This method will be more useful when we deal with a large number of random shares. When we want to track the owner of the shares, we just fold the share and the half portion of the share will be stacked with the remaining half portion of the share itself. This means shares contain visual information of both the image's tag as well as the secret. Distribution of both types of information on the share is tricky and researchable.

Points worth remembering:
Tagged visual cryptography is a method to address the problem of share handling in the case of a large number of shares.

Points worth remembering:
In TVC, shares contain information about both the image's tag as well as the secret.

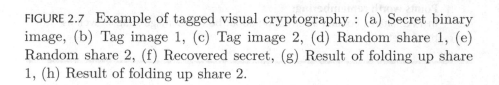

FIGURE 2.7 Example of tagged visual cryptography : (a) Secret binary image, (b) Tag image 1, (c) Tag image 2, (d) Random share 1, (e) Random share 2, (f) Recovered secret, (g) Result of folding up share 1, (h) Result of folding up share 2.

2.1.2.5 Friendly visual cryptography (FVC)

Randomness of the shares increases the vulnerability for cryptanalysis and may create confusion in share identification in case of a large number of participants. We can understand it by an analogy. Suppose we have generated fifty shares for an application using visual cryptography. Unfortunately an attacker has modified his share so that we cannot decrypt the exact secret. Since all shares are random in nature, it will be very difficult to identify the altered one. Hence, to minimize the difficulty for managing the huge number of shares, some meaningful information should be added. This meaningful information may be any additional information about shares or share holders like registered trademarks or any copyright logos, to prevent the mishandling or theft of the shares. All visual cryptography approaches which generate meaningful shares can be treated as FVC.

Figure 2.8 shows an example of FVC where the secret image is a halftone version of the Lena image. Here we can see that both shares have meaningful information rather than randomness. This meaningful information may be related to the share holder. Now if the number of shares is large, then we can simply identify the mishandled share as well as the owner of the share. In this example we have successfully addressed the problem of random shares but the problem of the increased dimension of the share still exists.

Points worth remembering:
Friendly Visual Cryptography (FVC) is visual cryptography with meaningful shares.

2.1.2.6 Size invariant visual cryptography

This type of visual cryptography is just an add-on to previously discussed approaches. All aforementioned approaches have the problem of increased size of shares that of secret image. This problem is because of pixel expansion m. Pixel expansion m is the number of subpixels by which one Secret Information Pixel (SIP) of the secret image is encoded. So m must be as small as possible because due to pixel expansion, there is a waste of the storage space and transmission time. We can understand it by an analogy. Suppose in case of telemedicine, a patient wants to send his medical report to an expert for online diagnosis. Due to security reasons (to avoid a man in the middle attack) the patient generates the shares of that secret medical image. If the size of the generated shares is larger than the actual secret image then it will take more bandwidth and transmission time which is highly undesirable. That is why the concept of size invariant visual cryptography came into the picture. Keeping the features of HVC and FVC as it is, this VC scheme maintains the size of shares the same as the secret image. Figure 2.9 shows an example of size

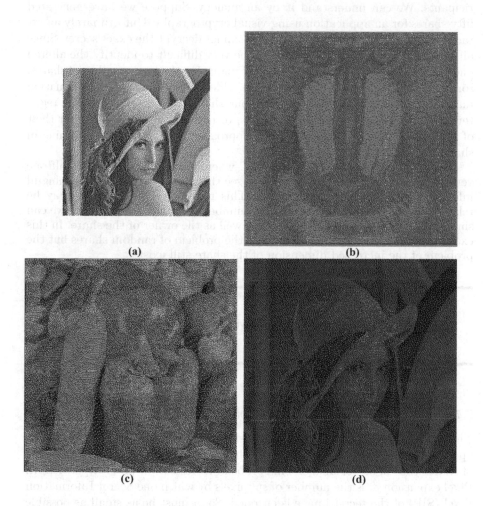

FIGURE 2.8 Example of Friendly visual cryptography: (a) Secret halftone image, (b) Meaningful share 1, (c) Meaningful share 2, (d) Recovered secret by stacking of share 1 and share 2 *(Courtesy of USC SIPI Image Database)*.

FIGURE 2.9 Example of size invariant visual cryptography: (a) Secret halftone image, (b) Meaningful Share 1, (c) Meaningful Share 2, (d) Recovered secret by stacking of share 1 and share 2.

invariant visual cryptography where the dimension of shares is the same as of the secret image and all shares are meaningful in nature. Here we can see that this example holds all properties of the aforesaid VC schemes where the secret image is halftone in nature (HVC property), shares are meaningful (FVC property) and the size of the share and secret are the same (size invariant VC).

The VC scheme shown in Figure 2.9 is more powerful in comparison to other discussed approaches. These properties are the basic requirements of VC which must be achieved during development of any visual cryptography approach.

Points worth remembering:
The dimension of secret and shares are identical in the case of the size invariant visual cryptography.

2.1.2.7 Progressive visual cryptography (PVC)

The traditional and threshold k-out-of-n or (k, n) visual cryptography is used to generate n shares from a secret image and distribute to n participants. The concept of threshold VC is based on the phenomenon of "All-or-Nothing" because the secret image can only be revealed when k or more shares are superimposed together and nothing is revealed if less than k shares are superimposed. In traditional visual cryptography, we cannot view the decoded image progressively because till now we assumed the applicability of VC only for the topmost secret applications. Hence, images must be visualized either in completely recovered form or nothing but noise (All or Nothing). Practically speaking, there are various applications in which traditional visual cryptography is hardly suitable. For example, in (n, k) VC if n and k are very large for any particular application then availability of k participants at a time is a major issue of concern. In this case we simply ignore the contribution of $k - 1$ participants. Hence, a new visual cryptography concept called progressive visual cryptography came into the picture in which clarity and contrast of a recovered secret is improved progressively by stacking more and more shares.

A simple analogy can be described to understand the concept behind Progressive Visual Cryptography (PVC). Let us consider a joint account of n account holders in a bank which provides multilevel services. These services will be available for the account holders with different levels of authenticity. We can understand the multilevel services as privileges at different levels, for example only balance inquiry can be done on the lower level services whereas upper level services are equipped with any kind of money transaction facilities. Level of authenticity is measured in terms of number of account holders. It increases with the number of contributing account holders in password decoding process. Here the bank password, generated by progressive visual cryptography is distributed in the form of shares to each account holder. No individual account holder has full access to the password as he has only a single share. Progressive stacking of two to n shares will give more and more visible passwords respectively. In the case of traditional k-out-of-n VSS, if because of some or the other reasons $n - k + 1$ share holders could not join the others in the bank, then the other $k - 1$ shareholders already present in the bank would not be able to use any of the bank facilities, because there is no clue of the password except noise by stacking less than k shares. One cannot decide the level of authenticity in traditional visual cryptography due to randomness of the recovered password (secret), obtained by the stacking of fewer than k shares. In case of PVC, stacking of more than one share out of n gives password hint instead of giving nothing as in traditional VC. In fact we cannot ignore the contribution of those shares which are present at the time of decoding. The number of password hints which goes towards the actual password decides the level of authenticity and hence eligible bank services for the same.

Figure 2.10 shows the secret halftone image which is going to be secretly shared. Six shares are generated by the progressive visual cryptography approach shown in Figure 2.10 (b) to (g). We can see here that all shares are random in nature as well as due to the pixel expansion problem, the dimension of the shares is larger than the secret image. Figure 2.11 shows the result of progressive stacking of the shares shown in Figure 2.10. Here we can see that Figures 2.11 (b) to (f) show the results of progressive stacking of shares two to n shares, respectively.

Points worth remembering:
In Progressive Visual Cryptography (PVC), clarity and contrast of the secret image is improved progressively by stacking more and more shares.

2.1.2.8 Progressive visual cryptography with meaningful shares without pixel expansion

We can combined the features of previously discussed visual cryptography approaches like friendly visual cryptography and size invariant visual cryptography with progressive visual cryptography in order to create an efficient approach. Figure 2.12 shows an example of PVC with unexpanded meaningful shares. Here we can see that all shares have some meaningful information which may be related to the share holder as well as all shares are of the same dimension as the secret image. This type of PVC is more applicable in various applications. In this example, the number of participants n is five. Figure 2.12

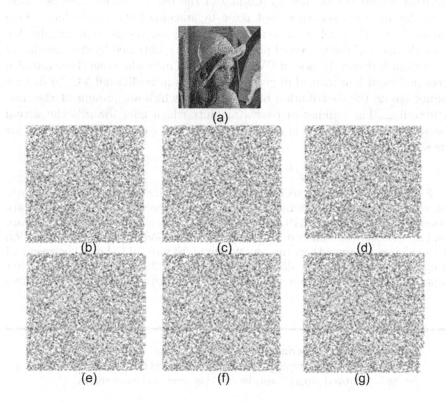

(a)

(b)   (c)   (d)

(e)   (f)   (g)

FIGURE 2.10 Secret image and shares of progressive visual cryptography: (a) Secret halftone image, (b)-(g) Random shares generated by the PVC approach.

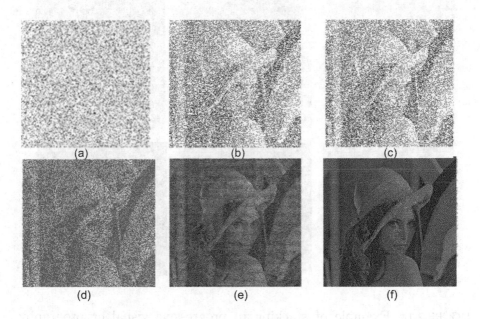

FIGURE 2.11 Example of stacking in progressive visual cryptography where $n = 6$: (a)-(f) Results of progressive stacking which were obtained after stacking one to n transparencies, respectively.

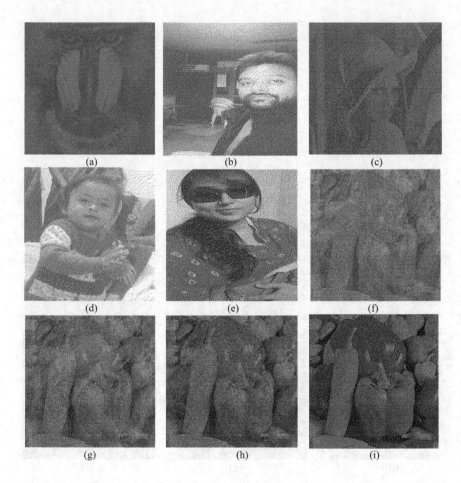

FIGURE 2.12 Example of stacking in progressive visual cryptography with meaningful shares where $n = 5$: (a)-(e) Meaningful shares for n share holders. (f)-(i) Results of progressive stacking which were obtained after stacking two to n transparencies, respectively. ((f) = $(a) + (b), (g) = (a) + (b) + (c), (h) = (a) + (b) + (c) + (d), (i) = (a) + (b) + (c) + (d) + (e)$ where '+'denotes a physical stacking operation).

(a)-(e) shows n meaningful shares. We can see that the dimension of all shares is identical to the original secret image which means this approach is a hybrid approach of FVC, PVC and size invariant VC. Figures (f)-(i) are the results of progressive stacking obtained after stacking two to n transparencies, respectively. We can understand the progressive stacking by the following relations: $(f) = (a) + (b), (g) = (a) + (b) + (c), (h) = (a) + (b) + (c) + (d), (i) = (a) + (b) + (c) + (d) + (e)$

where '+'denotes the physical stacking operation.

Points worth remembering:
One can make an efficient hybrid approach by combining the PVC, FVC and size invariant VC.

2.1.3   Dynamic Visual Cryptography (DVC)

There are many classes of visual cryptography which come in the category of Dynamic Visual Cryptography (DVC). There are many additional features like a color/gray image as secret, multiple secret sharing, etc, are added in the extended visual cryptography in order to develop dynamic visual cryptography approaches. These additional features are so unique and advanced that they divert DVC from the basic principles and findings of visual cryptography. For example, the very basic feature of Extended Visual Cryptography(EVC) approaches is indexcomputationless recovery. At the receiver end only just by stacking the predefined set of shares, one can recover the secret image. There are many other basic features of EVC like the type of secret image as binary/halftone only and the number of secret images to be shared. Any one or more of the basic features are compromised in all classes of dynamic visual cryptography.

Points worth remembering:
Dynamic visual cryptography approaches are nothing but extended VC with additional features.

2.1.3.1   Multitone/Continuous tone visual cryptography (MVC)

This approach is also called visual cryptography for gray/color images. Use of halftone (binary) images as secret is limited only to document (text) images, whereas in most of the applications we need to deal with natural and live secret images. A reduction in intensity range of secret pixels may lose sensitive

FIGURE 2.13 Example of multitone visual cryptography: (a) Multi-tone secret image, (b),(c) Random shares, (d) Recovered secret.

and important information of the secret image. We can understand it by an analogy: Let us consider a hospital associated with various scan and diagnosis centers. If unfortunately, an expert doctor who deals with medical images is unavailable then we need to transmit the medical images (MRI, CT Scan, X-Ray, etc and all are in JPEG or any two dimensional format) over the web to be examined by experts elsewhere. Medical images contain very sensitive information. Changes in intensity values of pixels in the report may result in addition/deletion of critical areas. This may mislead the experts and their diagnosis. This type of new security requirement laid the foundation of a new dimension in visual cryptography, i.e, visual cryptography with multi-tone images or MVC. One can peruse the Figure 2.13 for a better understanding of the MVC where (a) is multi-tone secret image, (b) and (c) are random shares and (d) is the recovered secret.

Points worth remembering:
There are many basic features such as contrast loss, computationless re-covery of a secret, binary image as a secret etc of extended and traditional VC which are compromised in dynamic visual cryptography.

Here we can observe that two main basic features of visual cryptography are lacking; the first one is computationless recovery and another one is a binary image as a secret. Computationless recovery means recovery by just stacking the shares which is only possible when the secret image is a binary/halftone. Unlike binary/halftone images where each intensity is made up of only a single bit, the intensities of multitone images are of eight/twenty-four bits in case of gray/color images. Therefore, a little bit computation at the receiver end is required in order to recover the secret image.

Points worth remembering:
All visual cryptography approaches which are made up for gray scale/color images are called multitone/continuous tone visual cryptography.

2.1.3.2   MVC with unexpanded meaningful shares

We can enhance the power and functionality of MVC by providing unexpanded meaningful shares. A share of MVC may be either binary or multitone. If we deal with binary shares for a multitone secret image, then shares will always be larger than the original secret. If we deal with multitone shares for multitone secret image, then the size of shares and the secret may be the same. We can understand it by an example. Let us assume for a moment that one intensity of a multitone secret image is 155 and we have generated its two corresponding share values 95 and 50 by any MVC method (discussed in the next chapter). If we write these share intensities in binary as 01011111 for 95 and 00110010 for 50, then each secret pixel will take eight times more space on the share. If the size of the secret is $m \times n$, then the size of the share will be $(m \times 8) \times n$ and each intensity will be denoted by eight binary bits in the share. The situation will be worse when we deal with a color image at that time because the size of the share will be $(m \times 24) \times n$ for the same secret. That is why in order to make unexpanded shares we always prefer multitone shares so that each intensity of secret takes only same space in each share. To make the shares meaningful we need to paste meaningful images on them so that they become no more random. This concept is related to imperceptibility, the spatial domain of an image and LSBs of the intensity which we will see in more detail in upcoming chapters. Figure 2.14 shows an example of MVC with meaningful and multitone shares.

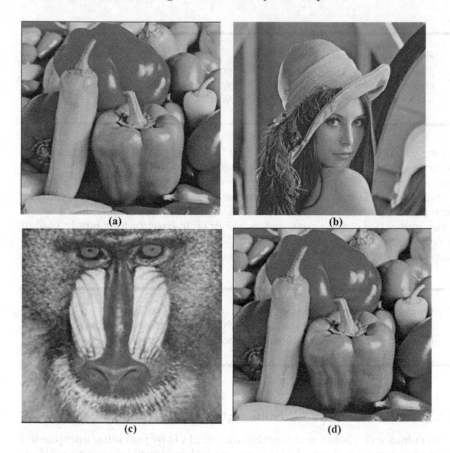

FIGURE 2.14 Example of multitone visual cryptography with multitone shares: (a) Multi-tone secret image, (b), (c) Meaningful multitone shares and (d) Recovered secret.

Points worth remembering:
An efficient approach can be proposed by merging MVC with FVC and size invariant visual cryptography.

2.1.3.3 Perfect recovery of the secret image in MVC

Contrast loss was the common problem in extended and traditional visual cryptography because of computationless recovery. We have already seen that MVC requires computation at the receiver end in order to decrypt the secret. That is why one can make such an algorithm which can perfectly decrypt the secret at the receiver end with the help of a little bit computation. In Figure 2.14 we can see that the original secret image (a) and recovered secret image (d) are completely identical to each other. This shows that we can remove the problem of contrast loss in MVC.

Points worth remembering:
MVC requires a little bit of computation at the receiver end, so the secret can be recovered without any contrast loss.

2.1.3.4 Visual cryptography with multiple secrets or multi secret sharing (MSS)

A very basic analogy can be considered to comprehend the necessity of multi secret sharing. Let us consider a case where one pair of persons have joint accounts in n different banks. n number of bank passwords are encrypted in the form of shares and distributed to both the account holders, so that no individual account holder cannot access the password. In case of a traditional or extended VC scheme, each account holder will have to carry n shares for n banks, since one share is embedded with only one bank password. Multi secret sharing can remove this constraint because a single share can be used to protect multiple secrets. Hence, if passwords are encrypted with MSS then each account holder will have to carry a single share for authentication in all the n banks because passwords for n banks can be revealed by stacking of two shares at n different angles.

One can look into Figure 2.15 for a better understanding of MSS where (c) and (d) are secret images recovered by placing the same set of shares at angle $0°$ and $90°$, respectively.

Points worth remembering:
In MSS, a greater number of secrets can be shared with the same set of shares.

2.1.3.5 Angle restriction problem in MSS

In the previous example, shares were rectangular in shape, which is why we were highly restricted to share only a limited number of shares. In a normal case, we have only four angles $(0°, 90°, 180°$ and $270°)$ to stack the rectangular shares, hence only four secrets can be secretly shared by rectangular shares.

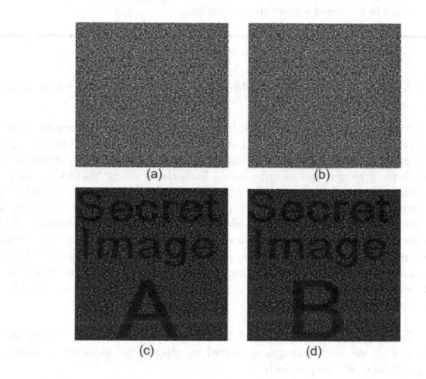

FIGURE 2.15 Example of multi secret sharing: (a) Random share 1, (b) Random share 2, (c) Recovered secret by placing share 1 and share 2 at 0° angle, (d) Recovered secret by placing share 1 and share 2 at 90° angle.

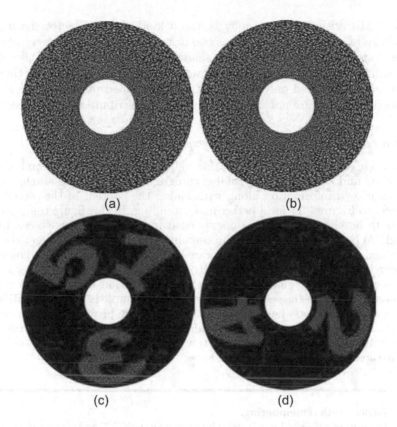

FIGURE 2.16 Example of Multi Secret Sharing with ring shares: (a) Random share 1, (b) Random share 2, (c) Recovered secret by placing share 1 and share 2 at 0° angle, (d)Recovered secret by placing share 1 and share 2 at 45° angle.

We can add more secrets if we use both sides of the shares for decryption. This type of approach is tough to implement. There is another way to make it a more flexible approach, by taking the shares in a ring or circle shape. In such a scenario we will have more angles to stack the shares, hence we can share more secrets at a time. One can refer to the Figure 2.16 for a better understanding of ring shares. Here we have many perspectives for rotation of the shares.

Points worth remembering:
Rectangular shares may have an angle restriction problem. To avoid it we may take shares in the form of a ring or circle.

Recursive visual cryptography is also a kind of multi secret sharing approach. Here, also, a share contains two or more secrets but recovery of those secrets can be done by shifting the shares on each other. In this case all secrets will be of different sizes as when we shift one share on another, then the secret will be revealed only because of the overlapped portion of the shares. All portions which are not overlapped will not contribute to secret recovery.

2.1.3.6 Multi secret sharing with unexpanded meaningful shares

We can create an efficient scheme for multi secret sharing by combining it with FVC and size invariant visual cryptography. In this case all shares will have their own information along with hiding the content of the secret. One can refer to Figure 2.17 for a better understanding. In this figure there are two secrets to be shared using two unexpanded meaningful shares shown by (c) and (d). When we directly superimpose both the shares, then we get the first secret and when the first share is rotated at 90° and then superimposed, we get secret 2. Here we can see that contrast loss of the recovered secret is very high because a single share is holding the information of its own as well as information for both the secrets. The presented example is for computationless recovery of the secrets, hence we cannot remove the problem of contrast loss in this case. But when we create an algorithm for multi secret sharing with computation (will see in the next section) then we can reduce the problem of contrast loss.

Points worth remembering:
Features of MSS can be enhanced by merging it with FVC and size invariant VC.

2.1.3.7 XOR-based visual cryptography

As previously discussed, all categories of visual cryptography suffer from the problems of pixel expansion, contrast loss in the recovered secret and random patterns on the shares until we provide the functionality explicitly. Computationless recovery always moves towards contrast loss as seen earlier. Perfect recovery of a secret image requires computation at the receiver end. To perfectly recover the secret images as well as for getting the unexpanded shares implicitly, XOR-based visual cryptography came into the picture. To make the shares meaningful, we again need to combine it with FVC.

An XOR logical operation has many properties which make it different from other logical operations. XOR gives maximum variations in output for two input values in comparison to other logical operators. It is also reversible in nature. We can see the following relations in an XOR operation:

Let P and Q be two images of the same size as $N_1 \times N_1$ and R be a ran-

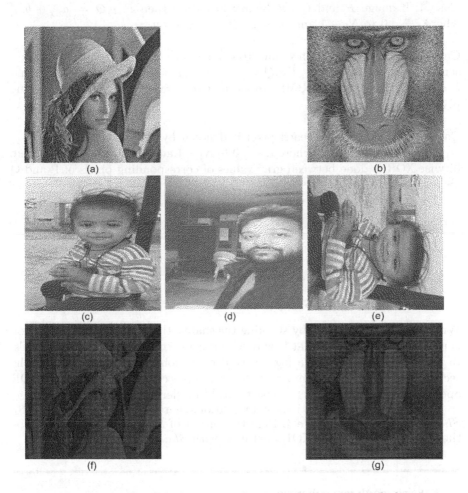

FIGURE 2.17 Example of multi secret sharing with unexpanded meaningful shares: (a) Halftone secret 1, (b) Halftone secret 2, (c) Unexpanded and meaningful share 1, (d) Unexpanded and meaningful share 2, (e) Share 1 at 90° angle, (f) Recovered secret by placing share 1 and share 2 at 0° angle, (g) Recovered secret by placing share 1 and share 2 at 90° angle.

dom matrix of the same size of P and Q. Let a and b be elements of P and Q respectively.

Case 1: If image P and Q are binary in nature then $P \oplus Q = [a_{i,j} \oplus b_{i,j}]$ where $i, j = 0 \ to \ N_1 - 1$ and $\oplus$ is an XOR operation.

Case 2: If image P and Q are multitone images then each pixel can be represented by eight bits. Hence, $P \oplus Q = [a_{i,j,k} \oplus b_{i,j,k}]$ where $i, j = 0 \ to \ N_1 - 1$ and $k = 0 \ to \ 7$. Here bitwise XOR is done between gray values of corresponding pixels of P and Q.

Case 3: For color images each pixel is denoted by 24 bits. Hence $P \oplus Q = [a_{i,j,kr,kg,kb} \oplus b_{i,j,kr,kg,kb}]$ where $i, j = 0 \ to \ N_1 - 1$ and $kr, kg, kb = 0 \ to \ 7$. Here bitwise XOR is done between gray values of corresponding pixels of P and Q for each color plane.

The followings are the valid XOR operations:

1. $P \oplus Q = Q \oplus P$

2. $P \oplus P = 0$

3. $P \oplus R$ is random where R is random.

4. $P \oplus Q = S$ implies $S \oplus Q = P$

When we do recovery just by stacking the shares, then a logical OR operation is performed indirectly. That is why we got some contrast loss in earlier VCs. In XOR-based VC due to a logical XOR operation, we perfectly recover the secret image. Here if n number of shares are present, then progressive XOR operations among all shares are performed in order to retrieve the secret.

Progressive XOR operations among various shares are defined as $Share \ 1 \oplus Share \ 2 \oplus Share \ 3 \oplus Share \ 4$. Here the output of $Share \ 1 \oplus Share \ 2$ will be the operand for the next XOR operation with $Share \ 3$ and so on.

Points worth remembering:
XOR-based VC is a computation-based visual cryptography approach which is used for perfect recovery of the secret .

2.1.3.8 *Hybrid approach with XOR-based VC, multitone VC, FVC, size invariant VC and multi secret sharing*

One can create a highly powerful approach by taking the contributions of all aforementioned VC approaches. This is still an open problem in visual cryptography. We can add various features in a single VC approach to enhance its functionality. We can make a type of VC approach which has unexpanded and meaningful shares for more than one multitone secret. The power of this approach can further be enhanced if we use XOR-based operations on shares to find out the secrets. XOR-based operations ensure the perfect recovery of the secrets. This hybrid approach almost denies the basic features and principles of traditional and extended visual cryptography as it has the capacity to share more than one multitone secret instead of a single binary/halftone share. It also requires computation at the receiver end. Secrets are recovered with full accuracy which is not possible in traditional and extended VC approaches. Figure 2.18 shows the example of a hybrid approach with XOR-based VC, multitone VC, FVC, size invariant VC and multi secret sharing. Here we can see that the same set of shares are used to share three multitone secrets and recovery of the secrets is done perfectly with the help of the XOR operation. All technical details behind this approach can be seen in the next chapters.

Points worth remembering:
Combining all the approaches of VC to provide better security is still under investigation.

2.1.3.9 Verifiable visual cryptography (VVC)

Security of a visual cryptography scheme is defined under the assumption that all the participants involved are trustworthy. If any participant acts as a cheater and tries to alter the share, then the recovered secret is compromised. It means shares are very sensitive objects and need to be protected in order to get an unaltered secret. When the number of participants involved in the VC process is large, then tracking the tampered share will be more difficult. To solve this problem, researchers have suggested a new dimension of VC i.e. Verifiable Visual Cryptography (VVC). One can easily understand the significance of VVC from Figure 2.19. Figure 2.19 shows a $(2, 2)$ threshold VC approach, incorporated with VVC where (a) is the secret image and (b) and (c) are protected shares, respectively. Alteration is done in share (b) by inserting Gaussian noise in it. If we decode the secret using the tampered share (d) and untampered share (e), we will not get the original secret. Since both shares still look fine hence no one can guess or suspect which share is altered by only using the human visual system (HVS). This problem will be more serious for (k, n) VC where n is large. Since in this example both shares are protected by VVC then one can easily find out which one is tampered. To track the alteration in shares, the algorithm related to VVC will be applied on shares to provide tamper localization as shown in Figure 2.19 (f) and (g). Figure 2.19 (f) does not show the pure black image like (g), instead here all white spot regions show the tampered regions.

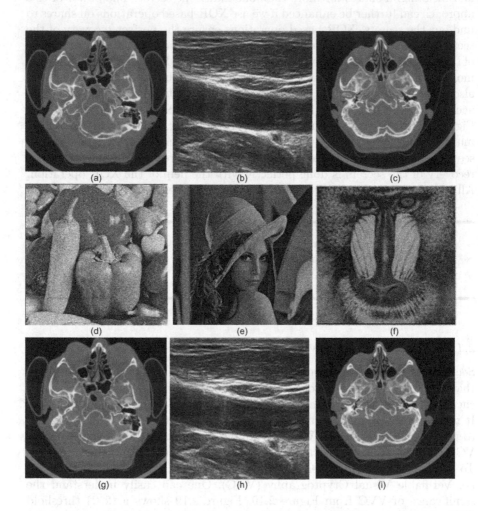

FIGURE 2.18 Hybrid approach with XOR-based VC, multitone VC, FVC, size invariant VC and multi secret sharing: (a) Multitone secret 1, (b) Multitone secret 2, (c) Multitone secret 3, (d)Unexpanded and multitone meaningful share 1, (e) Unexpanded and multitone meaningful share 2, (f) Unexpanded and multitone meaningful share 3, (g)-(i) Recovered secrets with progressive XOR operation(*Courtesy of Dr. Luca Saba, Radiology Department, University of Cagliari, Italy*).

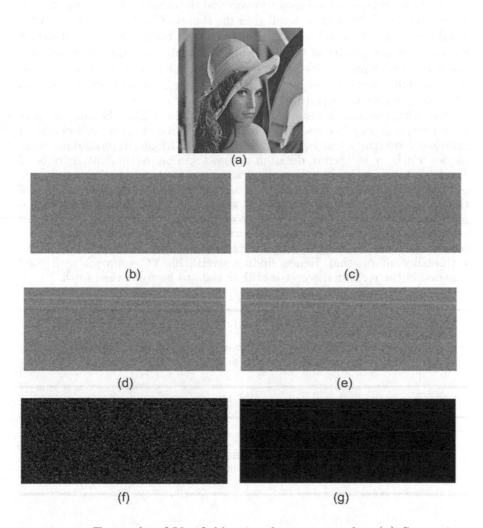

FIGURE 2.19 Example of Verifiable visual cryptography: (a) Secret image, (b), (c) Protected shares, (d), (e) Tampered shares, (f), (g) Shares with tamper localization.

2.1.3.10 Hybrid approach with VVC

Verifiable visual cryptography can be combined with any previously discussed VC approaches to provide better security. Till now we have seen two types of VC, one with computation-based recovery and the other without computation-based recovery. VVC can be applied in the shares of both types of VC. Verifiability can only be checked with computation at the receiver end. Hence, it increases the complexity of the algorithm a little bit. For example we can add verifiability with progressive visual cryptography or with multi secret sharing or with multitone visual cryptography, etc. There may be many variants of VC which can be developed by researchers.

Sometimes verifiable visual cryptography is also called cheating immune visual cryptography because of its immunity against cheating or intentional attacks. Verifiability can be introduced by using fragile watermarking techniques which we will see in detail in the next section. Verifiability may be of two types, one which just tells us that a given share is authentic/unauthentic or untampered/tampered and another one which tells us the exact location of the tampering. If we deal with the VC approaches which do not require computation for recovery, then verifiability will increase the contrast loss because each share will contain three things now. One is the secret information, the second one is the meaningful information of the share and the third one is the verifiability information. Hence, finding a verifiable VC approach with good contrast in the recovered secret is still an issue to be researched forth.

Points worth remembering:
VVC can be used to enhance the power of any existing VC approach by providing tamper detection capability to the shares.

Points worth remembering:
VVC can only be implemented with computation at the recovery end, therefore it may increase the complexity of the VC approach.

2.1.3.11 Random grid-based visual cryptography (RGVC)

A random grid can be defined as a transparency which comprises a two-dimensional array of pixels. Each pixel is either opaque or fully transparent in nature. A coin flip procedure is used to define the type of pixel (transparent or opaque). Hence, there are no relationships between the intensities of various pixels in the array. We can understand it by the phenomenon of light transmission. The transparent pixels will allow the light to pass from it whereas the opaque pixels stop it. The number of transparent pixels is probabilistically equal to that of the opaque pixels in a random grid as the probability of each pixel being opaque or transparent, is $1/2$. Hence, the average light transmission from a random grid is $1/2$. Let us consider that R is a random grid so the average light transmission denoted by $T(R) = \frac{1}{2}$. Random grid-based techniques can also be used to implement visual cryptography approaches in order to share any binary secret image. A random grid can be used to encrypt a secret binary picture into two random grids in which the areas containing information in the two grids are intercorrelated. When both random grids are superimposed together, the correlated areas will be resolved from the random background because of the difference in light transmission. In this way, the secret picture can be seen visually. In the case of RGVC, decoding is also done without any computation. This means this approach has the beauty of VC. How the random grids are generated for sharing a secret will be seen in the next chapter in full detail.

2.1.3.12 Hybrid approaches using RGVC

The main advantage of RGVC over other aforesaid VC approaches is that its shares are by default unexpanded. It means when we develop any VC approach using a random grid, it will be automatically size invariant in nature. But to enhance the efficiency and power of the RGVC we can combine it with other flavors of VC like FVC, MVC, etc. It is still an area under research to generate an efficient RGVC approach which has meaningful shares and which can share more than one secret at a time. The problem of contrast loss still exists in RGVC because it supports computationless recovery. The main challenge is to minimize the problem of contrast loss.

Points worth remembering:
Random grids are transparent with randomly distributed intensities of either white (transparent) or black (opaque).

Points worth remembering:
RGVC approaches generate unexpanded shares by default.

SUMMARY

- Visual cryptography schemes are broadly classified into three categories viz. *Traditional Visual Cryptography (TVC), Extended Visual Cryptography (EVC) and Dynamic Visual Cryptography (DVC).*

- All these approaches come under the category of *traditional VC* which follows the basis matrices in Shamir's approach.

- A *threshold VC scheme* is denoted as (k, n)-VCS, where $2 \leq k \leq n$.

- *Extended Visual Cryptography (EVC) approaches* slightly modify/extend the features of traditional visual cryptography by maintaining its basic principles.

- *Halftone images* are binary images with the illusion of gray scale images.

- Halftone images are more appropriate for visual cryptography in comparison to thresholded binary images because of their randomness.

- *Error Diffusion* is a well known method of halftoning which uses Floyd-Steinberg error filter.

- *Tagged Visual Cryptography (TVC)* is a method to address the problem of share handling in the case of a large number of shares.

- In TVC, shares contain information about both the image's tag as well as secret.

- Friendly Visual Cryptography (FVC) is visual cryptography with meaningful shares.

- The dimension of the secret and shares are identical in the the case of *size invariant visual cryptography.*

- In *Progressive Visual Cryptography (PVC)*, the clarity and contrast of the secret image are improved progressively by stacking more and more shares.

- *Dynamic Visual Cryptography (DVC)* approaches are nothing but an extended VC with additional features.

- There are many basic features like contrast loss, computationless recovery of the secret, binary image as the secret, etc in extended and traditional VC which are compromised in dynamic visual cryptography.

- All visual cryptography approaches which are made up for gray scale/color images are called *multitone/continuous tone visual cryptography.*

- *MVC* requires a little bit of computation at the receiver end so the secret can be recovered without any contrast loss.

- In *Multi Secret Sharing (MSS)*, one can share more secrets with the help of the same set of shares.
- *XOR-based VC* is a computation-based visual cryptography approach which is used for perfect recovery of the secret.
- *Verifiable Visual Cryptography(VVC)* can be used to enhance the power of any existing VC approach by providing a tamper detection capability to the shares.
- *Random grids* are transparency with randomly distributed intensities of either white(transparent) or black(opaque).
- *Random Grid-based Visual Cryptography(RGVC)* approaches generate unexpanded shares by default.

CHAPTER 3

Development of Visual Cryptography Approaches with Computationless Recovery of Secrets

CONTENTS

I N the previous chapter, we familiarized you with various visual cryptography approaches. They were categorized on the basis of their features. In this chapter we will see another categorization of visual cryptographic approaches on the basis of recovery of the secrets. As we know, traditional and extended visual cryptography approaches follow the very basic property of computationless recovery of the secrets, whereas most of the dynamic visual cryptography approaches violate this property of the VC and these approaches require computation at the receiver end in order to recover the secrets. Hence, VC approaches may now be categorized in two ways, that is, those VC approaches with computationless recovery of secrets and those VC approaches where some computation is required to recover the secret. In this chapter we will discuss the foundation of computationless VC schemes. Throughout the chapter, we will call these categories computationless (recovery of secrets without computation) and computation-based(Recovery of secrets with computation) VC schemes, respectively.

The **Chapter Learning Outcomes(CLO)** of this chapter are given below. After reading this chapter, readers will be able to:

1. Understand another categorization of visual cryptography.

2. Understand the technology behind the development of computationless VC schemes.

3. Develop his/her own hybrid and efficient visual cryptography algorithms as per the requirements.

4. Develop the MATLAB code for a basic implementation of the approaches.

3.1 COMPUTATIONLESS AND COMPUTATION-BASED VISUAL CRYPTOGRAPHY APPROACHES

All the traditional, extended and dynamic VC approaches where recovery of secrets is done without computation come under the category of compu-

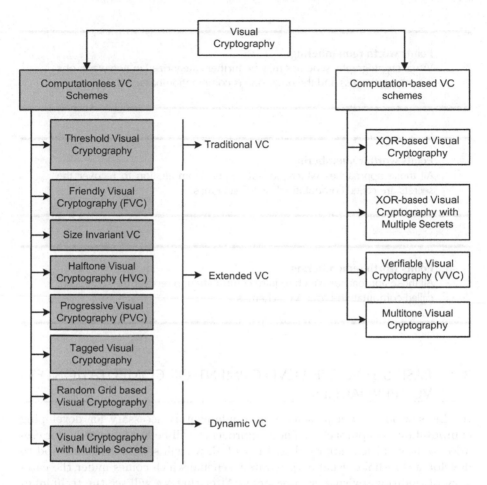

FIGURE 3.1 Categorization of computationless and computation-based VC schemes.

tationless recovery schemes. All the traditional, extended and dynamic VC approaches where recovery of secrets is obtained through some computation are called computation-based recovery schemes. One can understand this classification better by Figure 3.1. Here we see that many VC schemes fall into the category of computationless recovery approaches. A computationless VC scheme is the combination of all three VC categories (traditional, extended and dynamic). For example, a threshold VC scheme falls into the category of traditional VC whereas FVC, PVC, size invariant VC, tagged VC etc fall into the category of extended VC. At the same time, random grid-based VC and VC with multiple secrets fall into the category of dynamic VC schemes.

Points worth remembering:
Visual cryptography schemes may be further categorized in two ways, one is computationless and the other one is computation-based VC schemes.

Points worth remembering:
All those approaches which do not require computation to recover the secrets are called computationless VC schemes.

Points worth remembering:
All those approaches which require computation to recover the secrets are called computation-based VC schemes.

3.2 BASICS FOR THE DEVELOPMENT OF COMPUTATIONLESS VC APPROACHES

In this section we will present some fundamentals necessary for developing computationless approaches. These approaches will be the foundation for creation of new efficient approaches. First of all we will see the basic method to develop a threshold visual cryptography scheme which comes under the category of traditional visual cryptography. After that we will see the techniques to develop FVC, size invariant VC, etc and other hybrid approaches.

3.2.1 Development of Threshold Visual Cryptography

In order to develop a threshold visual cryptography scheme, we need to create basis matrices which will be used for preparing shares. In the previous chapter we saw that how basis matrices can be created using Shamir's approach. We will use that concept to understand the phenomenon behind threshold visual cryptography.

Example 3.1: Let us consider a threshold visual cryptography scheme for 3-out-of-3 VC. In this case when all three shares are stacked together, then only we can reveal the secret. Single or combinations of any two shares will not reveal the secret. For this 3-out-of-3 VC scheme, a set of participants W will be $\{W_0, W_1, W_2\}$ because there are only three participants. Hence, $\Gamma_{Qual} = \{\{W_1, W_2, W_3\}\}$ and $\Gamma_{Forb} =$

$\{\{W_1\}, \{W_2\}, \{W_3\}, \{W_2W_3\}, \{W_1W_3\}, \{W_1, W_2\}\}$.

Here $n = 3$, and set of participants $W = \{W_1, W_2, W_3\}$

$$\pi = \{\phi\}\{W_1W_2\}\{W_2W_3\}\{W_1W_3\}$$
$$\sigma = \{W_1\}\{W_2\}\{W_3\}\{W_1W_2W_3\}$$

Now according to Noar and Shamir's scheme, the size of the basis matrix will be $n \times 2^{n-1}$ i.e 3×4 where $m = 2^{n-1}$ for the given example.

$$S_{ij}^0 = 1, \text{ If } W_i \in \pi_j$$
$$\text{and}$$
$$S_{ij}^1 = 1, \text{ If } W_i \in \sigma_j$$

where $1 \leq i \leq n$ and $1 \leq j \leq 2^{n-1}$.
π and σ are used to construct the basis matrix S^0 and S^1, respectively.

$$S^0 = \begin{bmatrix} 0101 \\ 0110 \\ 0011 \end{bmatrix} \quad S^1 = \begin{bmatrix} 1001 \\ 0101 \\ 0011 \end{bmatrix}$$

The above example shows that in this case, a single binary pixel of the secret image will be encoded by four subpixels in each share. In this example, the number of shares is three. Now we can take the column permutation of both the matrices in order to create more options to encode. For example, S^0 may have various column permutations as shown below:

$$C^0 = \begin{bmatrix} 0101 \\ 0110 \\ 0011 \end{bmatrix} \begin{bmatrix} 1001 \\ 1010 \\ 0011 \end{bmatrix} \begin{bmatrix} 1010 \\ 1001 \\ 0011 \end{bmatrix} \dots \begin{bmatrix} 1001 \\ 1100 \\ 0101 \end{bmatrix}$$

Similarly, S^1 may have following column permutations:

$$C^1 = \begin{bmatrix} 1001 \\ 0101 \\ 0011 \end{bmatrix} \begin{bmatrix} 0101 \\ 1001 \\ 0011 \end{bmatrix} \begin{bmatrix} 1010 \\ 0110 \\ 0011 \end{bmatrix} \dots \begin{bmatrix} 1001 \\ 1010 \\ 0101 \end{bmatrix}$$

In order to encode any white pixel of the secret image we take any one matrix from C^0 and assign its subpixels from each row to the corresponding pixel of the share. In this example, one pixel of the secret image will be encoded by four subpixels and these four subpixels will be picked up from the row of the basis matrix.

One can understand the basic development of threshold VC by Figure 3.2. Here a binary secret image of size 2×2 is shared by the $(3, 3)$ threshold secret sharing scheme. According to Shamir's basis matrix creation approach, one

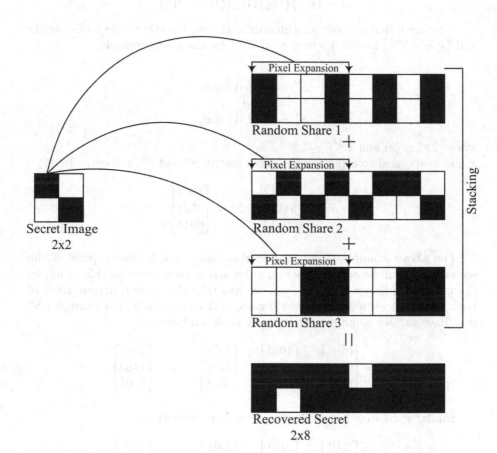

FIGURE 3.2 Example of the development process for threshold visual cryptography.

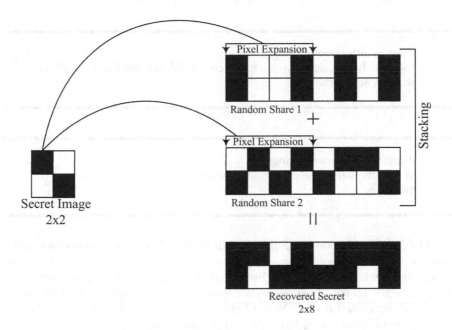

FIGURE 3.3 Example of a recovered secret with the forbidden set of shares.

pixel will be encoded by four subpixels as shown in the figure. These subpixels are chosen randomly from the row of any matrix C^j as in example 3.1. Whenever we stack all three shares, then we get black pixels with full accuracy but one black pixel will be recovered as four black subpixels. Similarly, a white pixel will be recovered with 25% contrast. This means one white pixel of the secret image is recovered with three black and one white subpixels. This scenario is called contrast loss. Since this is a threshold VC scheme, it means whenever we stack two or less than two shares, we will not get the secret as we can see in Figure 3.3. Here only two shares are stacked together which comes in the forbidden set Γ_{Forb}. We can see that we are unable to retrieve the secret image. Retrieval of the secret image with the help of an insufficient number of shares will be more complicated when the secret image is larger. In this case, the stacking operation will generate an ambiguous pattern of black and white subpixels for both secret black and white pixels.

This threshold VC scheme suffers from both major problems of the VC scheme i.e. the pixel expansion problem and a random pattern of the shares. We can see in Figure 3.2 that each single pixel of the secret image is expanded with four subpixels as well as that all the shares have meaningless information.

Points worth remembering:
The threshold visual cryptography scheme follows the basis matrix of Shamir's approach.

Points worth remembering:
In TVC, the size of the pixel expansion will increase with the increment of the shares.

The MATLAB code for 2-out-of-2 threshold VC is shown in codes 3.1, 3.2 and 3.3. Given threshold visual cryptography approach is applied on the secret image as shown in Figure 3.4. Here 2-out-of-2 VC scheme is implemented to share a secret image. Two shares are generated which are random in nature as well as expanded in size. When both shares are superimposed together we get the secret image back but with some contrast loss.

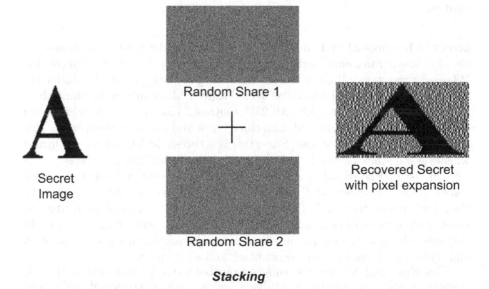

Secret Image

Random Share 1

+

Random Share 2

Stacking

Recovered Secret with pixel expansion

FIGURE 3.4 Example of a 2-out-of-2 threshold VC scheme.

MATLAB Code 3.1

```
%MATLAB Code 3.1
%MATLAB code for share generation for 2 out of 2 threshold VC.
inImg = imread('A.bmp');%Read Input Binary Secret Image
figure;imshow(inImg);title('Secret Image');
%Visual Cryptography
[share1, share2, share12] = VisCrypt(inImg);

%Outputs
figure;imshow(share1);title('Share 1');
figure;imshow(share2);title('Share 2');

figure;imshow(share12);title('Overlapping Share 1 & 2');

imwrite(share1,'Share1.bmp');
imwrite(share2,'Share2.bmp');
imwrite(share12,'Overlapped.bmp');
```

MATLAB Code 3.2

```
%MATLAB Code 3.2 for function VisCrypt()
%MATLAB code for share generation for 2 out of 2 threshold VC.
function [share1, share2, share12] = VisCrypt(inImg)

s = size(inImg);
share1 = zeros(s(1), (2 * s(2)));
share2 = zeros(s(1), (2 * s(2)));

%%White Pixel Processing
%White Pixel share combinations
disp('White Pixel Processing...');
s1a=[1 0];
s1b=[1 0];
[x y] = find(inImg == 1);
len = length(x);

for i=1:len
    a=x(i);b=y(i);
    pixShare=generateShare(s1a,s1b);
    share1((a),(2*b-1):(2*b))=pixShare(1,1:2);
    share2((a),(2*b-1):(2*b))=pixShare(2,1:2);
end

%Black Pixel Processing
%Black Pixel share combinations
disp('Black Pixel Processing...');
s0a=[1 0];
s0b=[0 1];
[x y] = find(inImg == 0);
len = length(x);

for i=1:len
    a=x(i);b=y(i);
    pixShare=generateShare(s0a,s0b);
    share1((a),(2*b-1):(2*b))=pixShare(1,1:2);
    share2((a),(2*b-1):(2*b))=pixShare(2,1:2);
end

share12=bitor(share1, share2);
share12 = share12;
disp('Share Generation Completed.');
```

3.2.2 Development of a Halftone Visual Cryptography (HVC) Scheme

A thresholded binary image is suitable to represent document (textual/numerical) images as we saw in the previous example. But in real life we have a number of variant image types. Most of the time we need to secretly share the natural or live images. In such cases, thresholded versions of these images will not be appealing visually when recovered, and from a security point of view, thresholded binary images for natural or live images are not recommended. That is why we need to make a secret image binary by using any existing halftone method before applying the VC algorithm. Halftone Visual Cryptography (HVC) is nothing but the VC approach using a halftone version of the secret image. HVC is the important phase for any advanced VC approach because nowadays all VC uses this technique to make the secret and shares more visually appealing. In the next sections, we will see how halftoning is important for VC.

MATLAB Code 3.3

```
%MATLAB Code 3.3 for function generateShare()
%MATLAB code for share generation for 2 out of 2 threshold VC.
function out = generateShare(a,b)

a1 = a(1);
a2 = a(2);
b1 = b(1);
b2 = b(2);

in = [a
      b];
out = zeros(size(in));
randNumber = floor(1.9*rand(1));

if (randNumber == 0)
    out = in;
elseif (randNumber == 1)
    a(1) = a2;
    a(2) = a1;
    b(1) = b2;
    b(2) = b1;
    out = [a
           b];
end
```

MATLAB Code 3.4

```
%MATLAB Code 3.4
%MATLAB code for Halftone image generation from its gray version using Floyd Steinberg
    Filter .
clc;clear all;close all;

%Getting Input
INPUT = 'lena.bmp'; %Take input any gray scale image.
in_img = imread(INPUT);
in_img= imresize(in_img,[256 256]);
imshow(in_img);title('Input Image');

%Halftone Image Conversion
halftone_img = floydHalftone(in_img);
figure;imshow(halftone_img);title('Floyd Halftoned Image');
imwrite(halftone_img,'lena_halftone.bmp');
```

MATLAB Code 3.5

```
%MATLAB Code 3.5
%MATLAB code for the function floydHalftone().
function outImg = floydHalftone(inImg)% Input image.
inImg = double(inImg);

[M,N] = size(inImg);
T = 127.5; %Threshold
y = inImg;
error = 0;

for rows = 1:M-1

    %Left Boundary of Image
    outImg(rows,1) =255*(y(rows,1)>=T);
    error = -outImg(rows,1) + y(rows,1);
    y(rows,1+1) = 7/16 * error + y(rows,1+1);
    y(rows+1,1+1) = 1/16 * error + y(rows+1,1+1);
    y(rows+1,1) = 5/16 * error + y(rows+1,1);

    for cols = 2:N-1
        %Center of Image
        outImg(rows,cols) =255*(y(rows,cols)>=T);
        error = -outImg(rows,cols) + y(rows,cols);
        y(rows,cols+1) = 7/16 * error + y(rows,cols+1);
        y(rows+1,cols+1) = 1/16 * error + y(rows+1,cols+1);
        y(rows+1,cols) = 5/16 * error + y(rows+1,cols);
        y(rows+1,cols-1) = 3/16 * error + y(rows+1,cols-1);
    end

    %Right Boundary of Image
    outImg(rows,N) =255*(y(rows,N)>=T);
    error = -outImg(rows,N) + y(rows,N);
    y(rows+1,N) = 5/16 * error + y(rows+1,N);
    y(rows+1,N-1) = 3/16 * error + y(rows+1,N-1);

end

%Bottom & Left Boundary of Image
rows = M;
outImg(rows,1) =255*(y(rows,1)>=T);
error = -outImg(rows,1) + y(rows,1);
y(rows,1+1) = 7/16 * error + y(rows,1+1);

%Bottom & Center of Image
for cols = 2:N-1
    outImg(rows,cols) =255*(y(rows,cols)>=T);
    error = -outImg(rows,cols) + y(rows,cols);
    y(rows,cols+1) = 7/16 * error + y(rows,cols+1);
end

%Thresholding
outImg(rows,N) =255*(y(rows,N)>=T);

outImg = im2bw(uint8(outImg));
```

The MATLAB code for halftone image generation using Floyd-Steinberg filter is shown in codes 3.4 and 3.5. This code is applied on a gray scale image of Lena as shown in Figure 3.5. Output of this code will be the halftone version of the input gray scale image of the same size.

FIGURE 3.5 Experimental result of image halftoning (using Floyd-Steinberg filter).

3.2.3 Development of a Friendly Visual Cryptography (FVC) Scheme

In this section we will see a very basic approach for making a Friendly Visual Cryptography (FVC). To make the FVC we need some cover images which will be displayed on the shares. To generate shares from these cover images we need some basis matrices which will be generated by the same method discussed earlier. We can understand this scheme by Figure 3.6. In this figure we have chosen the Lena image as the cover which will be displayed on the shares. This cover image may be gray or color, so first of all we need to covert it into binary. For binarization we may choose any halftoning method such as error diffusion so that the image looks more appealing in spite of conversion into binary. In the case of FVC, we have two major concerns. First one is the share must be meaningful in nature and it must show only the content of the cover, not the secret. The second concern is whenever we stack the shares, the cover information of the shares must be suppressed and it must show only the secret information. To achieve this goal we can take the negative of the halftone version of the cover image. When the original halftone and its negative are superimposed together, then it will show only an all black image. This approach is very basic. We can choose n different meaningful cover images for n shares but the above-mentioned two primary concerns must be satisfied. This suggested approach will only address the problem of random shares, not pixel expansion. Hence, the size of shares will be larger than the size of the secret. There will be a relation between the dimensions of both images (cover and secret). Let us assume that the size of the secret binary image is $m \times n$ and the size of the cover image is $M \times N$, then the size of shares will be $M \times N$. It means each secret pixel will have the pseudo pixel expansion of size $\frac{M \times N}{m \times n}$. For better understanding we can say that each secret pixel will be denoted by the halftone block of size $\frac{M}{m} \times \frac{N}{n}$. Now the question is why we are calling it pseudo pixel expansion? Actually, this halftone block will contain both types of information, for the cover image as well as the secret image. The information for secret pixel will be in expanded form by default as per the construction of the basis matrix. Hence, the cover image must include some additional space in order to hold the secret pixel.

The example shown in Figure 3.6 is an example of 2-out-of-2 FVC. Here, each secret pixel will be encoded by two subpixels according to Shamir's method. It means for each secret pixel, we have to randomly choose two positions for the halftone block each of size $\frac{M}{m} \times \frac{N}{n}$ of the halftone cover image. In Figure 3.6 the pixel locations for "B" and "A" are randomly chosen to embed the secret subpixels. The same positions must be selected in all the shares for secret subpixel embedding, so that whenever we superimpose all shares then information for the cover image will be cancelled out and only secret pixel information will be left. We can see in the figure that the same positions in the halftone block are selected in both the shares(original as well as negative) for secret subpixel embedding. These subpixels are selected row-wise as we have seen in previous threshold visual cryptography approaches. A

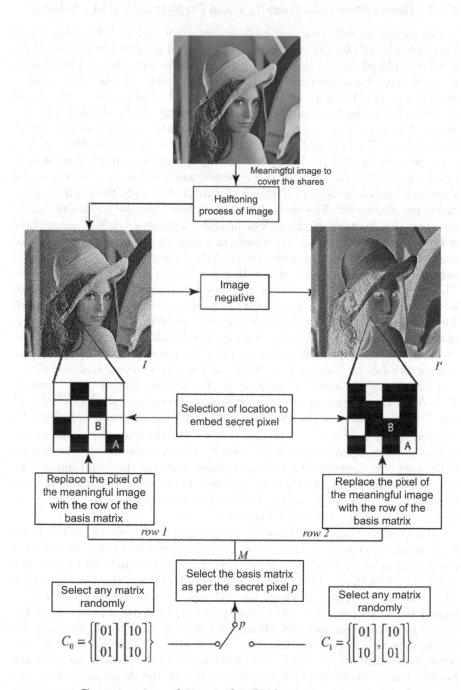

FIGURE 3.6 Construction of 2-out-of-2 FVC.

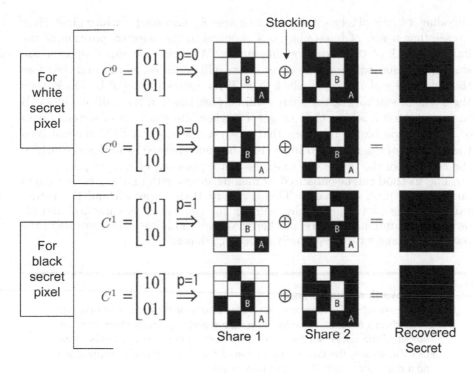

FIGURE 3.7 Replacement of the cover pixels with secret subpixels in the halftone block.

row will be selected from the basis matrix which is generated as per the pixel intensities (black or white). We know that we can make various combinations of basis matrices just by permuting its columns. Hence one can select a row of subpixels from any matrix, but then all rows will be selected from the same matrix for that pixel.

One can understand the replacement of cover pixels with secret subpixels in a halftone block from Figure 3.7. This example is also for a 2-out-of-2 VC scheme. Here, halftone blocks of size 4×4 are chosen for each pixel of the secret image from the selected cover image. Now we have sixteen pixel positions in the halftone block where we can embed the information related to secret subpixels. Whatever position in the halftone block we choose for embedding the secret subpixels in one share, the same position must be chosen in the other halftone block of the shares. We can change this position for each secret pixel to provide more security but for a single secret pixel, the position of embedding of secret subpixels in the halftone block of all shares must be the same. This is because of perfect stacking of secret subpixels and pixels of the cover image. The selected positions for embedding of secret subpixels are denoted as "A" and "B". We can see that the same positions are selected for the embedding in the corresponding set of shares for a secret pixel. Figure 3.7 shows two choices for encoding a black pixel as well as two choices for encoding a white pixel. Here we see that a row of basis matrix is assigned to the selected position of the halftone block of the corresponding shares. Whenever we superimpose these shares, information about the cover image will be cancelled out and therefore that location will appear as black. Now if the secret pixel will be black, then the result of stacking of the shares will be pure black in the halftone block. If the secret pixel is white, then we get a single white pixel out of sixteen pixels of the halftone block. It means that when we develop an FVC scheme then the contrast of the recovered image may be compromised as in this case where the contrast of the image is $\frac{1}{16}$ i.e. one white pixel out of sixteen pixels.

This method can be enhanced by making shares with multiple cover images and without pixel expansion. This is a very basic approach and the reader can create his own approach by utilizing the concept of this method and his own understanding. We have another challenge to improve the contrast of the recovered image with the visually appealing shares.

Points worth remembering:
In the case of FVC, there are two major issues of concern. The first one is that shares must be meaningful in nature and they must show only the content of the cover not the secret. The second concern is whenever we stack the shares, the cover information of the shares must be suppressed and it must show only the secret information.

The suggested friendly visual cryptography approach is applied on the secret image of Lena as shown in Figure 3.8. This approach is 2-out-of-2 visual cryptography approach, hence only two shares are generated. Here we can see that both shares are larger in size than the secret image because of halftone blocks. A complimentary pair of cover images are used so that they can cancel out their effect on the recovered secret image while stacking.

3.2.4 Development of Size Invariant Visual Cryptography

Till now we have seen visual cryptography schemes with meaningful shares. But these schemes suffer from the problem of pixel expansion. Now we will see the development of a method to implement size invariant visual cryptography. Preprocessing is necessary on a secret binary image before applying a visual cryptography scheme. First, we will see one of the preprocessing methods. Readers can determine their own method of preprocessing.

3.2.4.1 Preprocessing of secret image for size invariant visual cryptography

Preprocessing is necessary to create templates for the secret halftone image so that the dimensions of the preprocessed image and its corresponding secret

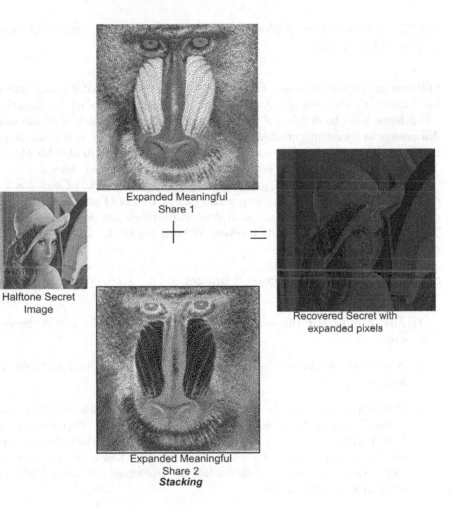

FIGURE 3.8 Example of 2-out-of-2 FVC.

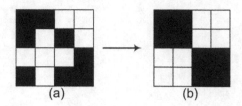

(a)   (b)

FIGURE 3.9   Conversion of blocks of the secret image into either full black or full white blocks.

halftone image are the same. First of all, a secret gray or color image will be binarized with the help of any halftoning algorithm. If a (n,n) size invariant VC scheme is to be developed, then n number of blank matrices of the same dimension as secret are created. It means if the size of the secret image I_{sec} is $M \times N$, then the dimension of all the n blank matrices will also be $M \times N$. These n blank matrices will be used to create n unexpanded shares.

The output preprocessed image will have only the blocks of size 2×2 of either complete black or complete white pixels. Blocks of all black pixels are denoted as B_{black} and blocks of complete white pixels are denoted as B_{white}. The value of M and N must be even. We can apply the following steps to achieve this objective.

Steps to create a preprocessed image:

1. First of all resize the halftone secret image so that M and N become even.

2. Now make the non-overlapping blocks of size 2×2 of pixels of the secret image.

3. Our objective is to make blocks either full of black pixels or full of white pixels. To do so, if the block contains three black pixels then convert the fourth white pixel into black and if three pixels are white then convert the fourth black pixel into white. One can refer to Figure 3.9 where we can see how the blocks of size 2×2 are converted into complete black or complete white blocks.

4. A problem will arise in converting the block to either completely black or white when there are two black and two white pixels in the block as shown in Figure 3.10. This is the decision-making point for whether that block should be converted into black or white. Many proposals are available but we can make our own proposal. Our objective will be to convert such a block to either in black or white in such a way that the structure of the image must be preserved. There must not be much difference be-

FIGURE 3.10 Example of a block with two black and two white pixels.

0	1	0	0
1	1	0	0
0	0	1	1
0	0	1	0

→

```
0  0
0  0
```

FIGURE 3.11 Example of a block with two black and two white pixels.

tween the original halftone image and the preprocessed halftone image in terms of imperceptibility and objective evaluation parameters.

5. One method to address this problem is the counting of neighbours. If a block has a majority of black neighbours then convert it into a black block, otherwise convert it into a white block as shown in Figure 3.11. Here we can see that the coloured block has the problem of two black and two white pixels. Now we count neighbours of pixels of the block. We see that the number of 0's are larger than 1's hence we convert all pixels of the block into 0. When the number of both intensities are the same then we can include one more bunch of neighbours. This is one method which can be used. Readers can determine their own efficient method for preprocessing.

6. We can apply replicate padding for all border pixels so that they can be processed without a problem. We can understand replicate padding from Figure 3.12. Here we see that all border pixels will be replicated but we can put 0 or 1 on the corner of most pixels as per our requirement.

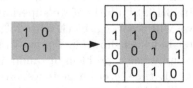

FIGURE 3.12 Example of a block with two black and two white pixels.

Points worth remembering:
Preprocessing is required in size invariant VC in order to keep the size of
the secret and shares the same.

Points worth remembering:
The objective of preprocessing is to create the predefined size clusters of
black and white pixels in the halftone image.

3.2.4.2 Size invariant share generation with the help of the preprocessed secret

Once we preprocess the secret image, we can create blank matrices of the same
size as the secret. For example if we are designing a VC scheme for 3-out-of-3
VC for the secret image of size, $M \times N$ then we need to generate three blank
matrices of size $M \times N$ for generating the shares. We will continue example
3.1 for a better understanding of this approach. The main concept behind this
approach is we have to treat each black or white block of size 2×2 of the secret
image as a single black or white pixel, respectively. After that we generate the
basis matrix for that black and white pixel. As in example 3.1, one secret pixel
is encoded with four subpixels. The same algorithm is applied here in order
to generate basis matrices for 3-out-of-3 VC. But in this case, all subpixels
are written in the rectangular shape of the block instead of the vector as
in the previous approach. One can understand the process of size invariant
share generation using the preprocessed secret and the blank matrices from
Figures 3.13 and 3.14. In this figure, we can see that a black block of size 2×2
of the secret is treated as a single black pixel. After that we generate three
shares with subpixel four using Shamir's basis matrix creation method. These
vectors of subpixels are now converted into the blocks of subpixels in the same
size as the secret block. Now we can see that all shares are of same size of
2 as the secret block. If we superimpose all the shares for the black block,
then we will recover an all black block of the same size as the original secret
block as shown in Figure 3.13. Similarly, if we superimpose all the shares for
the white block, then we will recover a block with 25% contrast of the same
size as the original secret block as shown in Figure 3.14. This suggested size
invariant visual cryptography has a problem of random shares. We can see in
the given examples that only the secret image is responsible for generating the
shares. There is no meaningful information present in shares. We can remove
this constraint by merging this approach with friendly visual cryptography.

The suggested size invariant visual cryptography approach is applied on

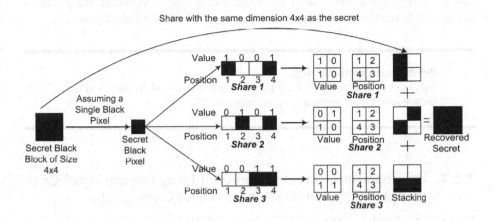

FIGURE 3.13 Size invariant share generation with the help of a prepro-cessed secret for a black pixel.

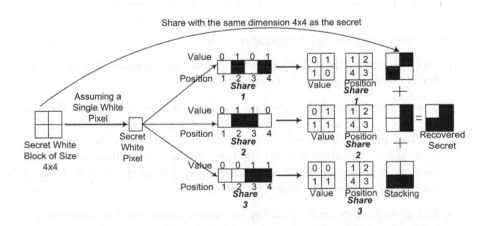

FIGURE 3.14 Size invariant share generation with the help of a prepro-cessed secret for a white pixel.

the image of Lena shown in Figure 3.15. This approach is the 3-out-of-3 visual cryptography approach. Here we can see that all three shares are random in nature, but of the same dimension as the secret image. When all these shares are stacked together, then the secret is recovered with contrast loss.

Points worth remembering:
Subpixels in the form of a vector are now rearranged in the form of a rectangular shape for creating size invariant VC.

3.2.5 Development of a Hybrid Approach Using Friendly Visual Cryptography (FVC) and Size Invariant Visual Cryptography

In order to develop a type of VC scheme which has unexpanded as well as meaningful shares, we need to merge both of the aforementioned techniques. Suppose we are developing a 4 out of 4 VC scheme with unexpanded meaningful shares. Then we need four meaningful cover images which will be displayed on the shares. Let us consider that the size of the secret image and all the cover images are $M \times N$. There may be many methods to carry out the same objective. Here we are discussing one of the methods but readers may develop their own effective method. This discussion is just to make them familiar with the development of these types of hybrid visual cryptography approaches with unexpanded meaningful shares. We have to follow the steps given below:

3.2.5.1 Steps for making a hybrid approach

1. Let us consider that S_1^{0011} & S_0^{0011} are the indications of two scenarios for two different pixels of the secret image.

2. One can understand the development method from Figure 3.16. Here we are explaining the development method for scenario S_1^{0011}. For a particular index of the secret image, S_1^{0011} represents the secret pixel value as 1 (shown as subscript) and the pixel values for all the corresponding indexes of four cover images (for meaningful shares) are 0, 0, 1 and 1 (shown as superscript), respectively.

3. In Figure 3.16 we are considering one secret image and four cover images of size 2×2. Figure 3.16 demonstrates that first of all, all of the secret and cover images are halftoned by any existing halftoning method.

4. After that, each image will be preprocessed. Due to preprocessing, all images including the secret and cover will have only the blocks of either four black or four white pixels. Now each block will be treated as a single

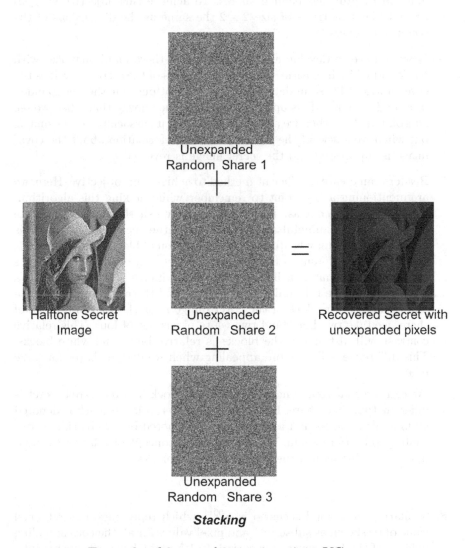

FIGURE 3.15 Example of 3-out-of-3 size invariant VC.

pixel. It means a black block of size 2×2 will be treated as a black pixel and a white block of size 2×2 will be treated as a white pixel.

5. Since we are going to develop 4-out-of-4 VC scheme, we need to create four meaningful unexpanded shares. To achieve this objective we generate 4 blank matrices of size 2×2 the same as the dimensions of the secret and covers.

6. Now we have to develop an algorithm to fill these blank matrices with black and white intensities based on the values of the secret as well as the cover images. These updated matrices are nothing but the unexpanded meaningful shares. This update must be in such a way that when we see an isolated share then we are only able to see its associated cover image, but when we stack all the shares, then the information about the cover image is suppressed and the secret image is revealed.

7. Readers can create an efficient method to achieve this objective. Here, we are mentioning a key point to remember while making this algorithm. We use relative contrast in order to distinguish the black and white intensities in meaningful shares as well as the recovered secret. This relative contrast may be different for indication of black and white pixels in shares and different in the recovered secret. For example, we can see in Figure 3.16 that if a block of the cover image is white, then in a block of share, we take only a single black and three white pixels out of four. Whereas if a block of the cover image is black, then in a block of share, we take two black and two white pixels out of four. This relative contrast will distinguish the blocks as relative black and white blocks. This difference will be more appealing when we design shares for large images.

We can see that representation of a black block in a recovered secret is different than in a share. In a recovered secret, a black block is denoted by four black pixels and a white block is denoted by two black and two white pixels. Here again, we are using the concept of relative contrast in order to distinguish the black and white blocks.

8. Similarly, the second scenario is S_0^{0011} which represents a secret pixel value of 0 (shown as subscript) and pixel values for all the corresponding indexes of the four cover images as 1, 0, 1 and 0 (shown as superscript), respectively. One can understand the generation of shares for given the scenario from Figure 3.17. Here, our secret block is white but we can see that we are unable to recover a complete white block. The recovered block has two white and two black pixels which ensures the recovered contrast is 50%.

9. One can conclude from the aforementioned examples that objective of

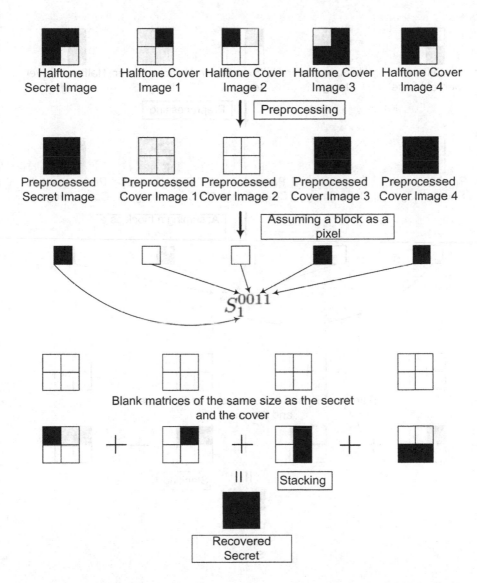

FIGURE 3.16 Development method for unexpanded meaningful shares (black block example).

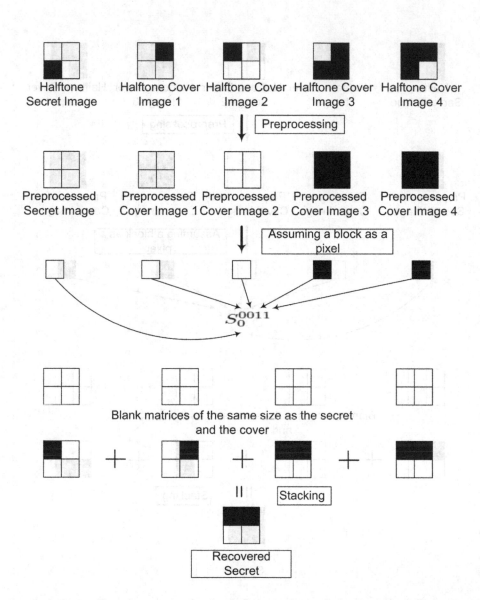

FIGURE 3.17 Development method for unexpanded meaningful shares (white block example).

the any proposed hybrid approach is to develop a type of technique which positions the black and white pixels within a block of size 2×2 in such a way that isolated shares exhibit the content of the cover image only. When all shares are superimposed together, then all these cover information must be suppressed in order to reveal the secret image.

Points worth remembering:
The existence of a hybrid approach is based on the concept of relative contrast difference.

Points worth remembering:
The difference between the number of black and white pixels within a block is responsible for making it a pseudo black or white block.

An experimental result is shown in Figure 3.18 which is obtained by applying the aforementioned hybrid approach on the secret image of Lena. This approach is 4-out-of-4 unexpanded meaningful shares. Here we can see that four shares are generated which are of the same dimensions as the secret as well as meaningful in nature. When all these shares are superimposed together, we can recover the secret image with contrast loss.

Points worth remembering:
In a hybrid approach, the arrangement of sub pixels for a secret pixel is decided by the secret pixel's as well as the corresponding cover image's intensities.

Points worth remembering:
In a block, subpixels are positioned in such a way that, an isolated share reveals the cover image, but a stacked share reveals the secret.

3.2.6 Development of Random Grid-Based Visual Cryptography

As we know, random grid-based VC has by default unexpanded shares. There are many approaches to do this. Here, we will see a very basic method to

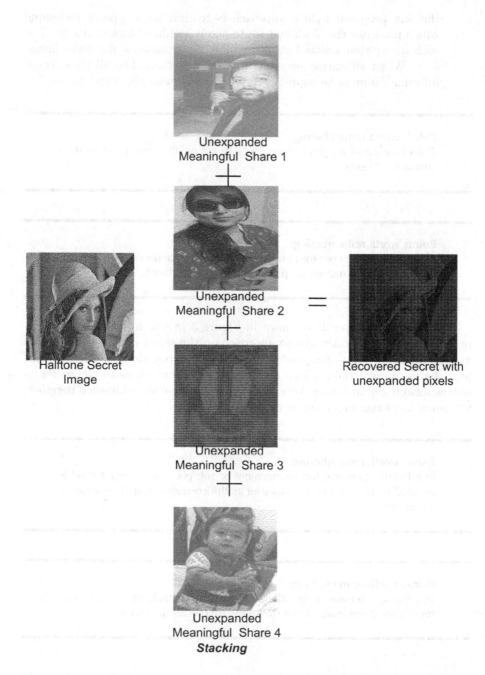

FIGURE 3.18 Example of 4-out-of-4 hybrid approach of VC with unexpanded meaningful shares.

generate shares using a random grid. Readers can create their own efficient approach.

3.2.6.1 Steps to generate shares using a random grid

1. Consider a secret image I of dimension $M \times N$. First of all, convert a gray/color secret image I into its halftone version using any existing approach.

2. Generate a random matrix R_1 of size $M \times N$. Here the probability density function of the randomization must be in such a way that it generates an approximately equal amount of black and white pixels.
$R_1(i,j) = RandomPixel(0,1)$ where $1 \leq i \leq M$ and $1 \leq j \leq N$. This random matrix R_1 will be treated as the first share. Here, we can see that the size of the share is unexpanded but the share is visually random.

3. Now we have to generate the second share. This share will be generated by using the value of the secret image I and R_1.

4. Let R_2 be the indication of the second share, then its values will be filled as follows:

5. If $I(i,j) = 0$ then $R_2(i,j) = R_1(i,j)$, otherwise $R_2(i,j) = R_1(\bar{i},j)$.

6. Now the generated R_1 and R_2 are binary in nature, hence stacking can be treated as logical OR operation and obeys the following truth table.

$$\begin{bmatrix} 0 \ OR \ 0 = 0 \\ 0 \ OR \ 1 = 1 \\ 1 \ OR \ 0 = 1 \\ 1 \ OR \ 1 = 1 \end{bmatrix}$$

In the process of generating random grids, we can see that black pixels (Intensity 1) are recovered with full accuracy, but the recovery probability of white pixels (Intensity 0) is not fixed. One can see the example of a random grid in Figure 3.19. Here we can see that first random share is absolutely random, generated by any random function, but the second share is dependent on the secret image and share 1 as mentioned in above approach. When both shares are superimposed together, then the black pixels are recovered with full accuracy whereas the occurrence of a white pixel is not fixed. The random grid-based approach is simple and easy to implement. One can make one's own efficient approach by adding the concept of friendly visual cryptography to a random grid to make meaningful shares. A MATLAB code is given in code 3.6 for share generation using random grid-based technique. One can see the experimental result of code 3.6 from random grid-based VC in Figure 3.20. The black pixels of the secret image are recovered with full accuracy. One can make an algorithm which can improve the contrast of the recovered image with meaningful shares. There is a considerable scope for future research in this field.

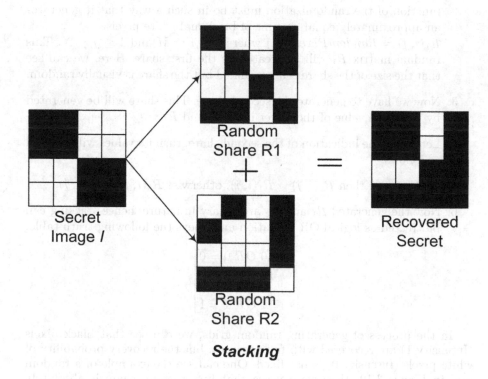

FIGURE 3.19 Development method for random grid-based VC approach.

Points worth remembering:
In random grid-based VC approach, shares are by default unexpanded.

MATLAB Code 3.6

```
%MATLAB Code 3.6
%MATLAB code for share generation using Random Grid.
I=imread('S.bmp');% Read halftone image
[r c]= size(I)
R1= rand(r,c);% Generation of first share
R2=rand(r,c); % Generated here but later on modified
for i= 1: r
    for j= 1:c
        if I(r,c)== 0
            R2(i,j)= R1(i,j);
        else
            R2(i,j)= R1(i,j);
        end
    end
end
figure, imshow(R1)
figure, imshow(R2)
Reco= bitor(R1,R2);% Stacking operation of both shares.
figure, imshow(Reco)% Reco is recovered secret.
```

3.2.7 Development of Visual Cryptography with Multiple Secrets

Till now all the approaches we have discussed are for sharing of a single secret. Now we will see how we can extend them to share more than one secret at a time. This type of scheme is little more challenging in comparison with other discussed approaches because of the dependency on more than one secret. For example, till now we were generating shares only on the basis of the pixels of a single secret if random shares were to be generated. The generation of shares become more complicated when we have to create meaningful shares. In this case shares are generated on the basis of a secret as well as the cover images (which are going to be displayed on the shares). Now the third scenario which we are going to deal with in this section is the same set of shares for multiple secrets. In this case shares will be generated on the basis of the pixel intensities of all the secrets as well as cover images (if we are talking about meaningful shares). This area is still under investigation. Here we are going to discuss a very trivial approach in order to enhance the reader's knowledge in this field so that readers can create their own effective approaches.

3.2.7.1 Steps to generate shares for multiple secrets

1. Since shares are to be rotated along their centers to recover the multiple secrets, first of all resize the secret images in dimension $M \times N$ where M and N are even and $M = N$.

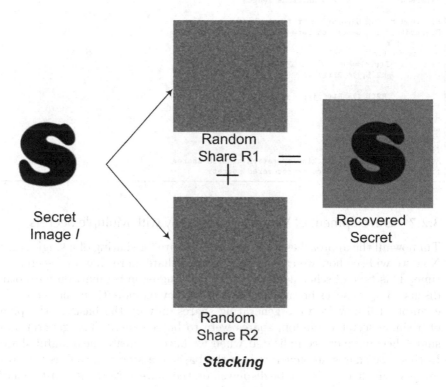

FIGURE 3.20 Experimental result for random grid-based VC approach.

2. Generate the halftone version of the secrets.

3. Suppose we are going to generate a 2-out-of-2 VC scheme, we need to create two blank matrices of size $M \times N$.

4. Virtually divide both the shares into four triangular blocks as shown in Figure 3.21. In this figure we can see that the division is done in triangular form and the blocks are numbered 0 to 3. Suppose we have two shares S_1 and S_2 and both are divided in the same manner. When both shares are superimposed directly (without any rotation), then blocks of both shares having an identical number will be overlapped. It means the 0th block of S_1 will be superimposed with the 0th block of S_2, the 1st block of S_1 will be overlapped with the 1st block of S_2 and so on.

5. If only one share of either S_1 or S_2 is rotated by 90° and both are superimposed, then k^{th} block of the non-rotated share will be overlapped with the $((k+3) mod\ 4)^{th}$ block of the rotated share where $k = 0..3$.

6. In Figure 3.21, we can see that pixels of all the blocks are numbered in a special manner. Due to this arrangement, when we rotate one share by holding another one, then the ith pixel of the kth block of a share will be stacked with the ith pixel of the $((k+3) mod\ 4)^{th}$ block.

7. Suppose we have two secret images I_1 and I_2 and we need to develop a 2-out-of-2 VC scheme to share both the secrets with the same set of shares. In order to recover the secret I_1, both shares S_1 and S_2 must be stacked directly but to recover I_2, share S_2 must be rotated at 90° before stacking with S_1.

8. If we want to create S_1 and S_2 then the ith intensity of the kth block will be decided by the ith intensity of the kth block of I_1 as well as the ith intensity of $((k+3) mod\ 4)^{th}$ block.

9. To achieve multiple secret sharing, secrets must be preprocessed first as done in size invariant VC so that we have four pixels in a block to hide both secrets. To check the efficiency of the VC approach we need to compare the recovered secrets with their corresponding preprocessed version. In Figure 3.22, we can see that both secret images are preprocessed first. The blocks of size 2×2 of both the shares are prepared with the blocks of both the secrets. This example is for unexpanded meaningless shares which is why the size of both shares is same as the secrets. In this case there will be more contrast loss because each share is holding information for both shares. If we develop a multi secret sharing approach with meaningful shares, then there may be more contrast loss because then the shares will hold information for the cover image also. In this example, the recovered image has only 25% contrast. Hence, an efficient approach for multi secret sharing with unexpanded meaningful shares is still an area under research.

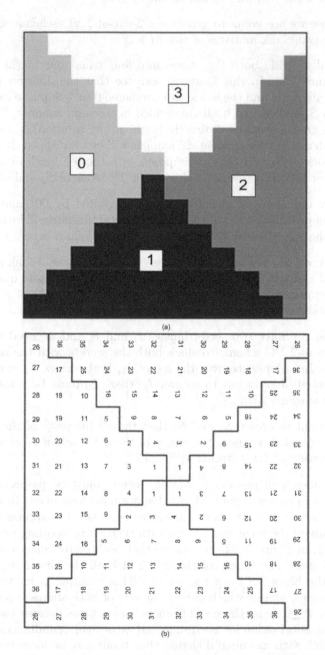

FIGURE 3.21 Division of shares for multiple secret sharing.

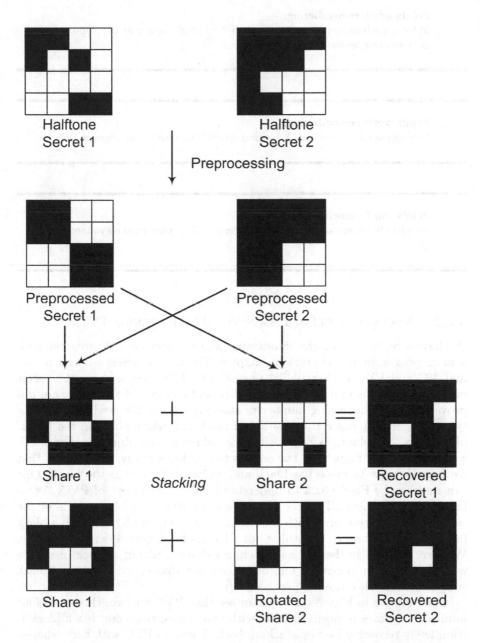

FIGURE 3.22 Development of 2-out-of-2 multiple secret sharing.

Points worth remembering:
In MSS, each share is dependent on all the secrets, hence its development is more complex than other VC schemes.

Points worth remembering:
Preprocessing of secrets with block division is required to implement MSS.

Points worth remembering:
MSS has the maximum contrast loss compared to other existing variants of VC.

3.2.8 Development of Progressive Visual Cryptography (PVC)

We have already studied the significance and requirements of progressive visual cryptography in the previous chapter. The main difference between normal VC and PVC is that unlike normal VC, PVC reconstructs the secret progressively. This means that the quality and contrast of the secret are improved with the increasing number of stacking shares. We need to develop a type of algorithm where the position of black and white pixels in the blocks of size 2×2 of shares is responsible for achieving the objective. For PVC, we again need to preprocess the secrets to develop size invariant VC so that we have options to move the black and white pixels within the block. One can refer to the Figure 3.23 to understand the basic concept of PVC. As we know, in PVC when all shares are stacked together, the black pixels of the secret image are recovered with accuracy. That is why we have chosen a black block of size 2×2 which is nothing but a block of a preprocessed secret image. We have to arrange the black and white subpixels within a block in such a way that with an increasing number of stacked shares, the number of black subpixels must also increase progressively.

For example in Figure 3.23 we can see that PVC for two, three and four numbers of shares is given. In PVC with two shares, the secret black block is completely recovered just by stacking both shares. In PVC with three shares, the secret black block is recovered progressively by adding more shares. When all three shares are stacked, we get the complete block. In PVC with four shares, the secret black block is recovered progressively by adding the shares. The complete block is recovered when all shares are stacked. Here we can see that by taking any set of incomplete shares, we are unable to recover the secret. When we deal with $n > 4$, then it may be possible that by stacking fewer shares, we may get some all black blocks. But it is not possible to get all black blocks for the whole secret image.

We can also create similar type of subpixel arrangement within a block for white subpixels. As we know, due to security reasons there is some contrast loss in the recovered images so we cannot recover the complete white block in spite of stacking all the shares. Figure 3.24 shows the arrangement of subpixels for a white secret block. Here we can see that after stacking of all shares, we get a pseudo white block with 25% contrast. We can see here that PVC for two, three and four numbers of shares is given. In PVC with two shares, the secret white block is recovered with 25% contrast just by stacking both shares. In PVC with three shares, the secret white block is recovered progressively by adding more shares. When all shares are stacked, we get a pseudo white block with 25% contrast. In PVC with four shares, the secret white block is recovered progressively by adding the shares. Block with 25% contrast is recovered when all shares are stacked. Here we can see that by stacking any set of incomplete shares we are unable to recover the secret. When we deal with $n \geq 4$ then it may be possible that by stacking fewer shares, we get the block as desired in the secret. Here, by stacking three shares, we get the same result as by stacking four shares. But, it does not mean that just by adding three shares we will get the complete exposure of the full secret image, since a secret image consists of many black and white blocks. So this is a rare case and we cannot get a completely recovered secret until the desired contrast of all black and white pixels is achieved.

To develop an efficient PVC, we need to develop a type of basis matrix creation algorithm which effectively redistributes the subpixels within the blocks of size 2×2 in order to achieve the objective of PVC. One can make a more efficient and effective algorithm by combining it with Friendly Visual Cryptography (FVC) and size invariant visual cryptography. In this way, we can get the unexpanded meaningful shares for PVC.

Points worth remembering:
PVC requires basis matrices which rearrange the pixels within a block in such a way that stacking of incomplete shares gives only a clue of the secret.

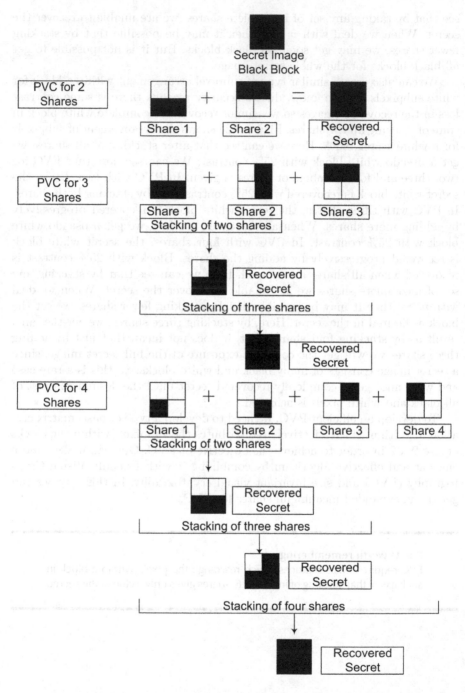

FIGURE 3.23 Development method of progressive visual cryptography to recover black block progressively.

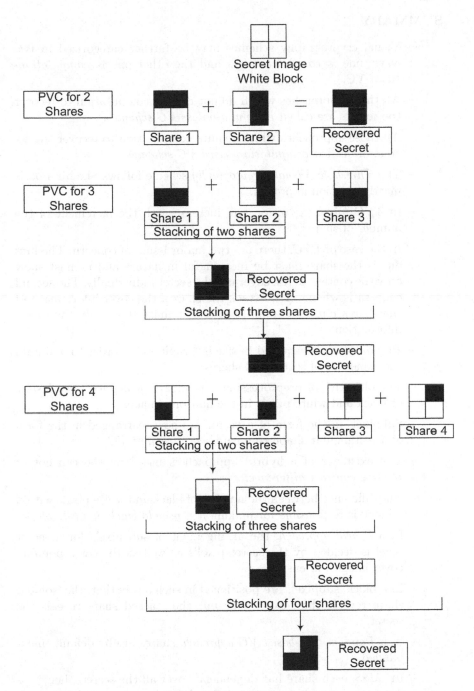

FIGURE 3.24 Development method of progressive visual cryptography to recover white block progressively.

SUMMARY

- Visual cryptography schemes may be further categorized in two ways, one is *computationless* and the other one is *computation-based* VC.

- All those approaches which do not require computation to recover the secrets are called *computationless VC schemes.*

- All those approaches which require computation to recover the secrets are called *computation-based VC schemes.*

- The *Threshold Visual Cryptography* scheme follows Shamir's basis matrix creation approach.

- In TVC, pixel expansion will increase with the increment of the number of shares.

- In the case of *FVC*, there are two major issues of concern. The first one is the share must be meaningful in nature and it must show only the content of the cover not the secret individually. The second concern is whenever we stack the shares, the cover information of the shares must be suppressed and it must show only the secret information.

- *Preprocessing* is required in size invariant VC in order to maintain the same size of secret and shares.

- The objective of preprocessing is to create *predefined size clusters* of black and white pixels in the halftone image.

- Subpixels in the form of a vector are now rearranged in the form of a rectangular shape to create *size invariant VC.*

- The existence of a hybrid approach is based on the concept of *relative contrast difference.*

- The difference between the number of black and white pixels within a block is responsible for making it a *pseudo black or white block.*

- In a *hybrid approach*, the arrangement of sub pixels for a secret pixel is decided by the secret pixel's as well as the corresponding cover image's intensities.

- In a block, subpixels are positioned in such a way that, the isolated share reveals the cover image but the stacked share reveals the secret.

- In a *Random grid-based VC approach*, shares are by default unexpanded.

- In *MSS*, each share has dependency over all the secrets, hence its development is complex than other VC schemes.

- Preprcessing of secrets with block division is required to implement MSS.

- *Multi Secret Sharing (MSS)*, has more contrast loss than other existing dimensions of VC.
- The *PVC algorithm* requires a type of basis matrix which rearranges the pixels within a block in such a way that stacking of incomplete shares gives only a clue to the secret.

CHAPTER 4

Development of Visual Cryptography Approaches with Computation-Based Recovery of Secrets

CONTENTS

I N the previous chapter, we studied various visual cryptography approaches which have computatioless recovery of secrets. In the case of dynamic visual cryptography, there are a number of visual cryptography approaches which have additional features that cannot be achieved by computationless approaches. To achieve these features, a little bit computation is required at the receiver end. As we are free to do computation, we have more freedom to avoid the various constrains of VC like contrast loss, binary secret images, etc. In this chapter, we will study the basics of the development of all those approaches which require computation at the receiver end in order to recover the secret image. These approaches are more flexible because we can add advanced features at any time. Due to computation, these approaches may have some computational complexities which must be considered during their development and implementation.

The **Chapter Learning Outcomes(CLO)** of this chapter are given below. After reading this chapter, readers will be able to:

1. Understand another categorization of visual cryptography.

2. Understand the mathematics and technology behind the development of computation-based VC schemes.

3. Develop their own efficient hybrid visual cryptography algorithms as per the requirements.

4. Develop the MATLAB code for the implementation of the approaches.

4.1 COMPUTATIONLESS AND COMPUTATION-BASED VISUAL CRYPTOGRAPHY APPROACHES

We have seen various classifications of VC based on the computation needed at the receiver end. In the previous chapter we became familiarized with all computationless VC approaches. We know that computationless VC contains the flavour of all three major types of VC such as traditional, extendible and dynamic. But computation-based VC contains only dynamic visual cryptography type approaches. One can understand the classification from Figure 4.1. Here we can see that there are four types of VC approaches in the category of computation-based VC which are XOR-based VC, XOR-based multiple secret sharing, verifiable visual cryptography and multitone visual cryptography. Here secrets are revealed at the receiver end by using calculations. These approaches are more flexible because we can mould them as per our requirements. These approaches also eliminate the many unnecessary encryption constraints of VC such as contrast loss of the recovered secret at the receiver end, restriction of the secret as binary, share and secret limitations, etc.

Points worth remembering:
The computation-based VC approach eliminates many unnecessary encryption constraints of VC such as contrast loss of the recovered secret at receiver end, binary image as secret, limitations of share and secret images etc.

4.1.1 Computation-based VC vs. Share Alignment Problem

In computationless VC approaches, we may face a very obvious problem of share alignment. As we know, in order to decrypt the secret, we must superimpose all the physical transparencies (shares) of the participants. Once these transparencies are stacked perfectly only then we can get the secret, otherwise nothing will be revealed or a very vague looking secret information will be revealed. Share alignment is the biggest problem in computationless VC approaches because this is the case where maximum human error is possible. Actually, there is a trade-off between the share alignment problem and pixel expansion. To create effective VC, we try to reduce pixel expansion but when pixel expansion is minimized, then the printed pixels on transparencies will be very small and hence difficult to align. If pixel expansion is large, then the pseudo size of the pixel will be large and can be easily aligned with the corresponding pixels of the other shares. So we can say that there is a trade-off

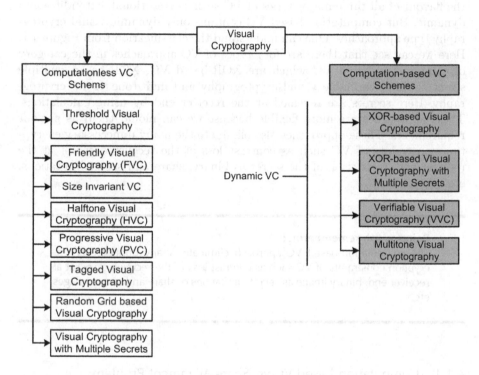

FIGURE 4.1 Categorization of computationless and computation-based VC schemes.

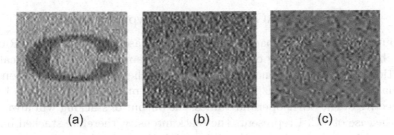

(a) (b) (c)

FIGURE 4.2 Problems due to stack alignment.

between the share alignment problem and pixel expansion. One can understand the seriousness of the problem of share alignment in computationless VC through Figure 4.2. Here we can see that Figure 4.2 (a) is the stacking result when all shares are stacked perfectly. Figure 4.2 (b) and (c) are the results when a single share is shifted by one and by two subpixels, respectively. One can conclude that the quality of the recovered secret degrades exponentially with the shifted pixels during stacking. Just by shifting two subpixels, we are unable to identify the secret image. This is the biggest problem in computationless VC approaches though it can be removed by computation-based VC.

Points worth remembering:
Share alignment is the biggest problem in computationless VC approaches.

Points worth remembering:
There is a trade-off between the share alignment problem and pixel expansion. When we increase pixel expansion, the share alignment problem is reduced and vice versa.

4.2 BASICS FOR THE DEVELOPMENT OF COMPUTATION-BASED VC APPROACHES

In this section, we will see the very basic methods used for developing computation-based approaches. These will work as the foundation for creation of our own efficient approach. First of all we will see a basic method to develop XOR-based visual cryptography scheme which comes under the category of dynamic visual cryptography. After that we will see the techniques to develop MVC, VVC, etc.

4.2.1 Development of XOR-Based Visual Cryptography

Computationless VC approaches are indirectly based on the logical OR operation, because the stacking operation of shares is nothing but the logical OR operation of their corresponding intensities. In the logical OR operation, the dominating operand value is 1. This means if any one of the operands is 1 then irrespective of the other operand values, the output of stacking will always be 1. In the case of VC, 1 represents the black intensity, therefore stacked images (recovered image) are darker than the original one in the case of computationless VC approaches. To avoid this contrast loss, we can switch to a logical XOR operation among the intensities of various shares. The XOR operation cannot be done directly just by stacking. A light weight computation or we could say a little bit of intervention of the processing unit is required to perform the XOR operation. In the XOR operation, the dominating operand is not 1 or 0, which is why there is no contrast loss. We can recover exactly the same secret that was shared. Since computers or a computing device is involved in the processing, we can say that there will be no share alignment problem in a XOR-based VC approach.

Points worth remembering:
Computationless VC is based on a logical OR operation whereas most of the time, computation-based VC is based on the XOR logical operation.

4.2.2 Basis Matrix Creation for the XOR-Based VC Approach

In computationless VC approaches, we have seen that generally the size of the basis matrix is $n \times m$ where n is the number of shares and m is the pixel expansion. In the case of XOR-based visual cryptography, there is no pixel expansion because we are doing computation. We can say that XOR-based VC has unexpanded shares by default. It means here, $m = 1$, i.e., one secret pixel is encoded by one pixel in each share. There are many approaches suggested in the literature for XOR-based VC. Here, we are going to demonstrate the development of a very basic approach for XOR-based VC. This is simply to improve the understanding of readers so that they can create their own efficient approaches.

Points worth remembering:
In XOR-based VC, shares are by default unexpanded.

Points worth remembering:
The size of the basis matrix in XOR-based VC is $t \times n$ where $t = \lceil \frac{n}{2} \rceil$.

First, we will see the development of the XOR-based VC approach with unexpanded but random shares. In order to design this approach, we need to make basis matrices for secret pixels 1 and 0 (since we are dealing with a binary secret image). In the case of XOR-based VC, the size of the basis matrix is $t \times n$ where $t = \lceil \frac{n}{2} \rceil$ and n is the number of participants. In computationless VC, each row of the basis matrix is dedicated to a share, but here each column of a selected single row is dedicated to a share. We can understand it through the example given. First of all, we will need to know how to make a basis matrix for XOR-based VC. We use a very basic property of the XOR operation in our consideration in order to create the basis matrix. In any XOR operation when the number 1s are odd, then it will return 1 as the output otherwise 0 as the output. For example, $1 \oplus 1 \oplus 0 \oplus 1 = 1$, so here the number of 1s are odd and $1 \oplus 1 \oplus 0 \oplus 0 = 0$, and here the number of 1s is even. Actually, this is the successive XOR operation where the output of two consecutive operands will be operated with the next operand and so on. We can refer to the truth table of XOR operations given below:

$$\begin{bmatrix} 0 & \oplus & 0 & = & 0 \\ 0 & \oplus & 1 & = & 1 \\ 1 & \oplus & 0 & = & 1 \\ 1 & \oplus & 1 & = & 0 \end{bmatrix}$$

Now we need to create two basis matrices, one for intensity 1 and another one for 0. The size of both matrices will be $t \times n$. Let $I = \begin{bmatrix} 1 & 1 \\ 0 & 0 \end{bmatrix}$ be a secret image of size 2×2, $n = 3$ and the basis matrices for 0 and 1 are S^0 and S^1, respectively. As we know, $t = \lceil \frac{n}{2} \rceil$ i.e. $t = 2$, hence the size of both matrices would be of 2×3. The matrices S^0 and S^1 are as follows:

$$S^0 = \begin{bmatrix} 0 & 0 & 0 \\ 0 & 1 & 1 \end{bmatrix} \quad S^1 = \begin{bmatrix} 0 & 0 & 1 \\ 1 & 1 & 1 \end{bmatrix}$$

Here we can see that S^0 contains an even number of 1s in each row whereas S^1 contains an odd number of 1s in each row. To encode the secret pixel 0, any one of the row can be selected from S^0 randomly and its elements will be assigned to corresponding pixels of successive shares. Similarly, to encode the secret pixel 1, any one of the row can be selected from S^1 randomly and its elements will be assigned to corresponding pixels of shares. We can take the various column permutations of the basis matrices to enhance the options to encode so that the security can be improved.

$$C^0 = \left\{ \begin{bmatrix} 0 & 0 & 0 \\ 0 & 1 & 1 \end{bmatrix}, \begin{bmatrix} 0 & 0 & 0 \\ 1 & 0 & 1 \end{bmatrix}, \begin{bmatrix} 0 & 0 & 0 \\ 1 & 1 & 0 \end{bmatrix} \right\}$$

$$C^1 = \left\{ \begin{bmatrix} 0 & 0 & 1 \\ 1 & 1 & 1 \end{bmatrix}, \begin{bmatrix} 1 & 0 & 0 \\ 1 & 1 & 1 \end{bmatrix}, \begin{bmatrix} 0 & 1 & 0 \\ 1 & 1 & 1 \end{bmatrix} \right\}$$

Let the column permuted matrices be C^0 and C^1. Here we can see that any one matrix can be chosen from C^0 or C^1 to encode 0 and 1, respectively. Now we need to generate three random shares each of size 2×2. The first pixel intensity of the secret image $I(1,1) = 1$ will be encoded with the basis matrix S^1. Suppose we choose the first row randomly from S^1 i.e. [0 0 1], now the successive elements of this row will be assigned at the $(1,1)^{th}$ positions of the respective shares. Similarly, $I(2,1) = 0$ will be encoded by the basis matrix S^0. Suppose we have chosen the second row randomly from S^0 i.e. [0 1 1], now its element will be assigned to the $(2,1)^{th}$ positions of respective share's intensity of each share. In this way, we can assign the value to each pixel intensity of the share. If we superimpose (XOR) all the shares then all the $(i,j)^{th}$ pixels of each share will be XORed together. It means for the $(1,1)^{th}$ location recovered (XORed) value will be $0 \oplus 0 \oplus 1 = 1$ and for the $(2,1)^{th}$ location recovered (XORed) value will be $0 \oplus 1 \oplus 1 = 0$. We can conclude that we are XORing all the elements of rows to recover the secret. Since in S^0, the number of 1s is even, it returns 0 as the output of XOR and in S^1, the number of 1s is odd, so it returns 1 as the output of XOR. We can understand it from Figure 4.3. Here we can see how the column values of the row of basis matrices are used to create shares. The same process is repeated for all pixels of the secret image. The rows are chosen randomly from the basis matrix to provide security.

The aforementioned suggested algorithm is implemented using MATLAB code as shown in code 4.1. The experimental result of this code is shown in Figure 4.4 where three random unexpanded shares are generated. Here we are taking the halftone version of the Lena image as the secret and we perfectly recover the secret shown in (e) at the receiver end.

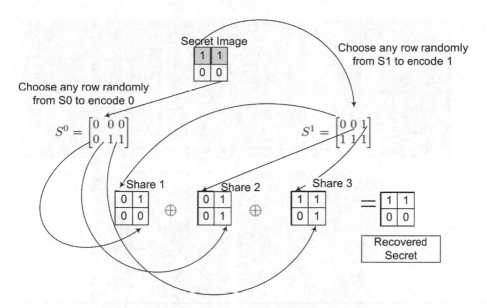

FIGURE 4.3 Generation of random unexpanded shares using XOR-based VC.

MATLAB Code 4.1

```
%MATLAB Code 4.1
% This MATLAB code is for demonstration of three random share
%generation using XOR-based VC.
clc
clear all
close all
I=imread('lena_halftone.bmp');
[r c]= size(I);
bm0=[0 0 0;0 1 1];
bm1=[0 0 1;1 1 1];
for i= 1: r
    for j= 1:c
if I(i,j)== 0
        randNumber = round(rand(1))+1;
        Share1(i,j)=bm0(randNumber,1);
        Share2(i,j)=bm0(randNumber,2);
        Share3(i,j)=bm0(randNumber,3);
else
        randNumber = round(rand(1)+1);
        Share1(i,j)=bm1(randNumber,1);
        Share2(i,j)=bm1(randNumber,2);
        Share3(i,j)=bm1(randNumber,3);
end
    end
end
figure,imshow(Share1)
figure,imshow(Share1)
figure,imshow(Share1)
Secret= bitxor(Share1,bitxor(Share2,Share3));
figure,imshow(Secret)
```

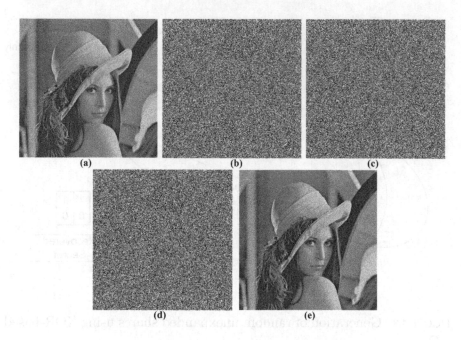

(a) (b) (c)
(d) (e)

FIGURE 4.4 Experimental result for generation of random unexpanded shares using XOR-based VC.

4.2.3 XOR-Based VC with Unexpanded Meaningful Shares

In the previous approach, we understood the basic development of XOR-based VC with unexpanded shares. Now we add a new feature in the shares that is meaningfulness. Since in the case of meaningful shares, the share holds both kinds of information viz secret information as well as information to be present on that share i.e. information related to cover image, we need to process the shares on a block basis. Here we are suggesting one approach but readers can create their own efficient approaches.

4.2.3.1 Steps to develop XOR-based VC with unexpanded meaningful shares

1. In order to create meaningful shares for XOR-based VC, first we need to create random shares using the method explained earlier.

2. Now we need to collect n halftone images of the same size as the secret which will be displayed on the shares. Let us call them cover images.

3. Preprocess the cover images as discussed in earlier chapters in order to create the blocks of size 2×2 either of complete black or complete white pixels.

4. Now we need to assign the one or more pixels (white or black) of the

blocks of the preprocessed cover image to the corresponding blocks of random shares in such a way that the total count of 1s in the cumulative $(i, j)^{th}$ positions in each share must not be changed.

5. The number of the pixels of cover image on share decides the clarity of the meaningful shares.

One can better understand the concept of the development of meaningful share generation using XOR-based VC from Figure 4.5. Here we can see that a halftone secret image is taken of size 4×4 and three meaningful shares are generated of the same size as 4×4. First, all cover images are preprocessed and a single black/white block of the cover image is treated as a single black-/white pixel. Now we randomly choose a position in a block of the random share to embed that cover pixel. In Figure 4.5, the index $(1, 1)$ is chosen in first block. If the embedded pixel is the same as the already present pixel in the random share, then do nothing, otherwise change the pixel value. Due to the change in the intensity of the random block, the counting of 1s will be changed which is not desirable. Hence, we need to change the intensity of the same position in other random shares. We can choose such a position in the 2×2 block for cover information embedding which requires limited changes. Figure 4.5 demonstrates embedding in a single block only. The same thing will happen for the entire image. This is a very basic approach and readers can create their own effective technique for making meaningful shares.

Points worth remembering:
One can share n number of multitone secrets using n number of shares using XOR-based MSS.

4.2.4   Development of Multitone Visual Cryptography

In the previous section, we discussed one of the development methods for XOR-based visual cryptography approach for binary images. A XOR-based method can handle perfect recovery of the secrets. In this section we will see a secret sharing method for multitone secrets. Here, we may use various logical operations and arithmetic operations in order to recover the secret, hence it comes under the category of computation-based VC. We use bitwise operation on bits of the pixels of the secret image. Here, we discuss one of the basic methods for the development of multitone visual cryptography. Readers can find inspiration from this and develop their own efficient methods.

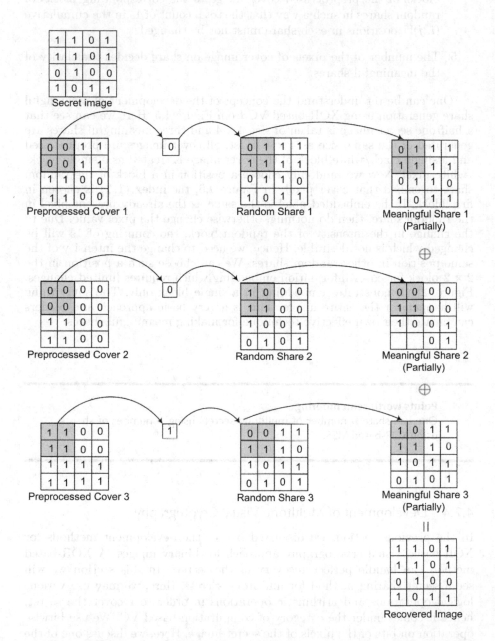

FIGURE 4.5 Example of XOR-based VC with unexpanded meaningful shares.

4.2.4.1 Steps to develop multitone visual cryptography with random shares

1. Let $I = \begin{bmatrix} 128 & 36 \\ 29 & 10 \end{bmatrix}$ be a multitone secret image of size 2×2. First of all convert this image into its binary version i.e. $\begin{bmatrix} 10000000 & 00100100 \\ 00011101 & 00001010 \end{bmatrix}$.

2. Now we have various options to generate random shares for secret image I. For example, we can make eight shares by taking the corresponding bit planes from each pixel and then encrypt those shares with any secret key. The eight bit planes are $\begin{bmatrix} 1 & 0 \\ 0 & 0 \end{bmatrix}, \begin{bmatrix} 0 & 0 \\ 0 & 0 \end{bmatrix}, \begin{bmatrix} 0 & 1 \\ 0 & 0 \end{bmatrix}, \begin{bmatrix} 0 & 0 \\ 1 & 0 \end{bmatrix}, \begin{bmatrix} 0 & 0 \\ 1 & 1 \end{bmatrix}$, $\begin{bmatrix} 0 & 1 \\ 1 & 0 \end{bmatrix}, \begin{bmatrix} 0 & 0 \\ 0 & 1 \end{bmatrix}, \begin{bmatrix} 0 & 0 \\ 1 & 0 \end{bmatrix}$.

3. All these shares may be shuffled or encrypted with a secret key and at the receiver end again, we use the same symmetric key to reshuffle the bit plane matrices.

4. At the end, these all the bit planes may be combined to create intensity of the eight bit pixel.

5. This is the simplest method which generates unexpanded random shares. In order to reduce the number of shares, we can pick four pairs of the bits from the each pixel of I to generate four shares. After that, we can apply Shamir's basis matrix algorithm to the shares to make them more secure.

6. Here, we are free to make any number of shares by using various combinations of the bit planes of the original intensity of secret image I. To recover the secret, we just have to make a reverse of the share generation approach. Since it is a computation-based approach, we are free to use any kind of arithmetic or logical operations. One thing we must acknowledge is that with the increments in the computation during share generation, the complexity will also increase and hence secret recovery will be more complex as well.

MATLAB Code 4.2

```
%MATLAB Code 4.2
% This MATLAB code is for demonstration of share generation of multitone secret using bit
    plane slicing
clc
clear all
close all
%Bit slicing
i=imread('Lena.bmp'); % Read the image
[r c]= size(i);
s1=input('Enter the Secret Key1 for Authentication');
  rand('state',s1);
Sm= logical(round(rand(r,c)));
share1=bitxor(Sm,logical(bitget(i,1)));
share2=bitxor(Sm,logical(bitget(i,2)));
share3=bitxor(Sm,logical(bitget(i,3)));
share4=bitxor(Sm,logical(bitget(i,4)));
share5=bitxor(Sm,logical(bitget(i,5)));
share6=bitxor(Sm,logical(bitget(i,6)));
share7=bitxor(Sm,logical(bitget(i,7)));
share8=bitxor(Sm,logical(bitget(i,8)));
```

Points worth remembering:
To implement multitone visual cryptography we are free to use any kind of arithmetic and logical operations

Points worth remembering:
Pixels of multitone secrets must be converted first into their binary form.

One can see the experimental result shown in Figure 4.6 generated by the code given 4.2. Here, eight unexpanded random shares are generated by using the bit plane slicing method. Figure 4.6 (a) and (j) are original and recovered secrets, respectively, whereas Figure 4.6 (b) to (i) are random shares generated by bit plane slicing method. We can simply decrypt the bit plane from the shares by using the same secret key, and by using the "bitset" function in MATLAB, we can simply reconstruct the original intensity.

Points worth remembering:
In a multitone VC approach, we are free to take any number of shares.

In the above mentioned approach, we saw a very basic approach for random share generation. Now we will see how to generate meaningful shares for multitone images. Actually, here we are presenting a very basic approach for

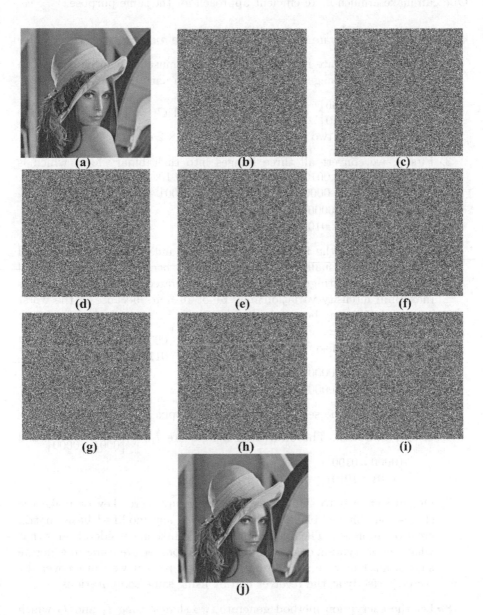

FIGURE 4.6 Example of multitone visual cryptography with bit plane slicing.

meaningful share generation for multitone secrets to clarify our understanding. One can make a much more efficient approach for the same purpose.

4.2.4.2 Steps to generate meaningful shares for a multitone secret

1. In order to generate meaningful shares, we must have cover images. In this case, cover images must be multitone in nature.

2. Let $I = \begin{bmatrix} 128 & 36 \\ 29 & 10 \end{bmatrix}$, $C_1 = \begin{bmatrix} 50 & 12 \\ 129 & 110 \end{bmatrix}$ and $C_2 = \begin{bmatrix} 150 & 112 \\ 12 & 210 \end{bmatrix}$ be the secret image and two cover images of size 2×2 each.

3. Firstly, we convert all three images into their binary form which is $I = \begin{bmatrix} 10000000 & 00100100 \\ 00011101 & 00001010 \end{bmatrix}$, $C_1 = \begin{bmatrix} 00110010 & 00001100 \\ 10000001 & 01101110 \end{bmatrix}$ and $C_2 = \begin{bmatrix} 10010110 & 01110000 \\ 00001100 & 11010010 \end{bmatrix}$.

4. If we just make the first four LSBs of C_1 and C_2 as 0, then we still will be able to visually interpret C_1 and C_2 because our human visual system cannot distinguish the difference generated by these changes. The maximum intensity variation may be of 16 units because in the worst case 1111 (15) may be changed to 0000(0).

5. Now the modified $C_1 = \begin{bmatrix} 00110000 & 00000000 \\ 10000000 & 01100000 \end{bmatrix}$ and $C_2 = \begin{bmatrix} 10010000 & 01110000 \\ 00000000 & 11010000 \end{bmatrix}$.

6. Now, bifurcate the secret image I's binary pixels into two equal parts of four bits each. The bifurcated images are $I_1 = \begin{bmatrix} 1000 & 0010 \\ 0001 & 0000 \end{bmatrix}$ and $I_2 = \begin{bmatrix} 0000 & 0100 \\ 1101 & 1010 \end{bmatrix}$.

7. Our objective is to shuffle the bits using any secret key or make the shares four bit per pixel using Shamir's or any modified basis matrix creation approach. The main thing we must take in consideration is that whatever encryption or randomization methods we are using to generate shares, must be reversible in nature. This is so that we can recover the secret perfectly at the receiver end by using same computations.

8. Let the encryption method generate two shares using I_1 and I_2 which are $S_1 = \begin{bmatrix} 1100 & 1010 \\ 0111 & 0110 \end{bmatrix}$ and $S_2 = \begin{bmatrix} 0100 & 0101 \\ 0001 & 0010 \end{bmatrix}$.

9. In the end, in order to generate meaningful shares, we just embed these four bits of each position into the first four LSBs of the corresponding pixels of C_1 and C_2.

10. So the final meaningful shares are $Sm_1 = \begin{bmatrix} 00111100 & 00001010 \\ 10000111 & 01100110 \end{bmatrix}$ and $Sm_2 = \begin{bmatrix} 10010100 & 01110101 \\ 00000001 & 11010010 \end{bmatrix}$.

11. To recover the secret at the receiver end, extract the first four LSBs of both meaningful shares. After that decrypt or reshuffle it using the same symmetric key. Finally, merge both of the four bits to make a complete eight bits of intensity per pixel. This generated image will be the final recovered image.

One can refer to Figure 4.7, which is the experimental result of the afore-mentioned approach for unexpanded meaningful share generation for multi-tone secret image. Here, Figure 4.7 (a) is the multitone secret image and (b) and (c) are two random shares generated by any encryption method applied on the first four LSBs of pixels in (a). Figure 4.7 (d) and (e) are two meaningful shares generated by embedding into the first four LSBs of two cover images. (f) is the recovered secret image. We can see that we are able to recover the secret image perfectly at the receiver end.

4.2.5 Development of XOR-Based Multi Secret Sharing Approach for Multitone Secrets

In computationless VC approaches multi secret sharing was restricted to lim-ited shares. At the same time the quality of the recovered image was highly degraded with the increasing number of secrets. If computation-based multi secret sharing is implemented using the XOR-based method, then we can elim-inate both the aforementioned problems. One can share n number of secrets with n number of shares and all secrets can be recovered at the receiver end with full accuracy. Unlike a computationless multi secret sharing approach, multitone secrets can be shared using a computation-based multi secret shar-ing approach. In this case all shares will also be multitoned.

4.2.5.1 Steps to develop XOR-based multi secret sharing approach for mul-titone secrets with random shares

Before understanding the development of a XOR-based multi secret sharing approach, we must understand the XOR operation for multitone images. Our suggested method consists of two types of operations, namely, the XOR opera-tion between two multitone images and the progressive XOR operation among various multitone images.

Let P and Q be two multitone images of the same size as $N \times N$ and R be a random matrix of the same size as of P and Q. Let a and b denote the elements of P and Q, respectively.

Case 1: For multitone images, each pixel value is represented by eight bits.

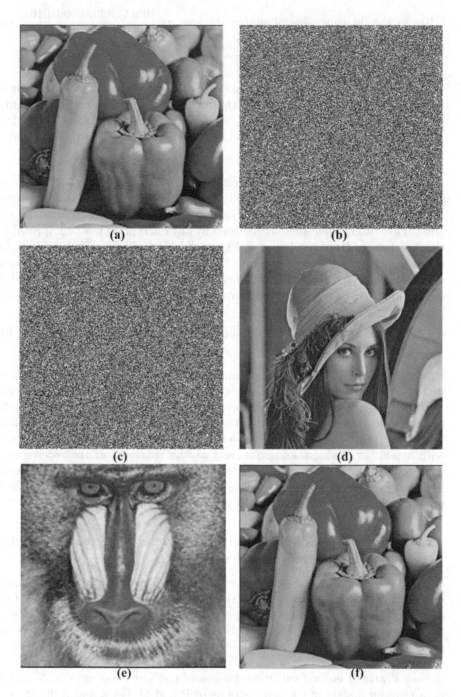

FIGURE 4.7 Experimental result of multitone visual cryptography.

Hence, $P \oplus Q = [a_{i,j,k} \oplus b_{i,j,k}]$ where $i, j = 0$ to $N - 1$ and $k = 0$ to 7. Here, bitwise XOR is done between corresponding gray level pixels of P and Q.

Case 2: For color images each pixel is denoted by 24 bits. Hence $P \oplus Q = [a_{i,j,kr,kg,kb} \oplus b_{i,j,kr,kg,kb}]$ where $i, j = 0$ to $N - 1$ and $kr, kg, kb = 0$ to 7. Here bitwise XOR is done between corresponding gray level pixels of P and Q for each color plane.

The XOR operation also satisfies the following operations which are beneficial for our method:

1. $P \oplus Q = Q \oplus P$

2. $P \oplus P = 0$

3. $P \oplus R$ is random where R is random.

4. $P \oplus Q = S$ implies $S \oplus Q = P$ (The most important property for multi secret sharing)

The progressive XOR operation among various multitone secret or share images P,Q,R,S is defined as $P \oplus Q \oplus R \oplus S$ and is performed similarly. Here the output of $P \oplus Q$ will be the operand for the next XOR operation with R and so on.

Here we note a very basic method for XOR-based multi secret sharing. This is not the only approach, readers can create their own efficient approach.

Points worth remembering:
Reversibility is a very important property of the XOR operation, which is used in multi secret sharing.

---

1. Suppose we have two secret multitone images of size 2×2 that are $I_{sec_1} = \begin{bmatrix} 20 & 50 \\ 128 & 75 \end{bmatrix}$ and $I_{sec_2} = \begin{bmatrix} 56 & 16 \\ 175 & 85 \end{bmatrix}$ which are going to be secretly shared simultaneously.

2. First, the we need to generate a master random share S_m of the same size 2×2 using a secret key. Let $S_m = \begin{bmatrix} 50 & 77 \\ 100 & 105 \end{bmatrix}$ be that master random share.

3. Now we will generate two random shares I_{sec_1} and I_{sec_2} using S_m. For I_{sec_1}, the bits of master share S_m will be circularly left rotated by one bit and the resultant matrix will be $\begin{bmatrix} 100 & 154 \\ 200 & 210 \end{bmatrix}$, hence the first random matrix $R_1 = \begin{bmatrix} 20 & 50 \\ 128 & 75 \end{bmatrix} \oplus \begin{bmatrix} 100 & 154 \\ 200 & 210 \end{bmatrix}$, that is, the first share $S_1 = R_1 = \begin{bmatrix} 112 & 168 \\ 72 & 153 \end{bmatrix}$.

4. To generate a second share, the master share S_m will be circularly left rotated by two bits and the resultant matrix will be $\begin{bmatrix} 200 & 53 \\ 145 & 165 \end{bmatrix}$, hence, second random matrix $R_2 = \begin{bmatrix} 56 & 16 \\ 175 & 85 \end{bmatrix} \oplus \begin{bmatrix} 200 & 53 \\ 145 & 165 \end{bmatrix}$, that is, $R_2 = \begin{bmatrix} 240 & 37 \\ 62 & 240 \end{bmatrix}$.

5. Share 2 can be calculated by R_1 and share S_1. $S_2 = S_1 \oplus R_2$ that is $S_2 = \begin{bmatrix} 112 & 168 \\ 72 & 153 \end{bmatrix} \oplus \begin{bmatrix} 240 & 37 \\ 62 & 240 \end{bmatrix}$ or $S_2 = \begin{bmatrix} 128 & 141 \\ 118 & 105 \end{bmatrix}$.

6. Now we have three shares S_1, S_2 and S_m. There is no need to send S_m because it can be generated at the receiver end by the same secret key by which it has been generated at the sender side.

7. At the receiver side, in order to get I_{sec_1} first we calculate the one bit circular shifted version of S_m denoted by S_m^1 and then calculate $I_{sec_1} = S_1 \oplus S_m^1$.

8. Once we get the I_{sec_1} then we calculate I_{sec_2}. For that first we determine $R_2 = S_2 \oplus S_1$.

9. Then we determine the two bit-shifted version of S_m denoted by S_m^2 and calculate $I_{sec_2} = R_2 \oplus S_m^2$.

MATLAB codes 4.3 and 4.4 are the implementation of the above mentioned procedure. Here we are sharing two multitone secrets using two shares and one master share.

MATLAB Code 4.3

```
%MATLAB Code 4.3
%MATLAB code for multitone muilt secret sharing with random shares.
I1= imread('Med1.bmp');
I2= imread('Med2.PNG');
figure, imshow(I1)
figure, imshow(I2)
I1=imresize(I1,[256 256]);
I2=imresize(I2,[256 256]);
s1=input('Enter the Secret Key1 for Authentication');
    rand('state',s1);
Sm= floor(rand(256,256)*256);
Sm= uint8(Sm);
figure, imshow(Sm)
Sm1= circshift(Sm,1);
Sm2= circshift(Sm,2);
R1= bitxor(I1,Sm1);
Share1= R1;
R2= bitxor(I2,Sm2);
Share2= bitxor(Share1,R2);
figure,imshow(Share1)
imwrite(Share1,'share1.jpg')
figure,imshow(Share2)
imwrite(Share2,'share2.jpg')
```

MATLAB Code 4.4

```
%MATLAB Code 4.4
%MATLAB code for recovery of multitone muilt secret sharing with random shares.
Share1= imread('share1.jpg');
Share2= imread('share2.jpg');
figure, imshow(Share1)
figure, imshow(Share2)
s1=input('Enter the Secret Key1 for Authentication');
    rand('state',s1);
Sm= floor(rand(256,256)*256);
Sm= uint8(Sm);
figure, imshow(Sm)
Sm1= circshift(Sm,1);
Sm2= circshift(Sm,2);
I1= bitxor(Share1,Sm1);
R2= bitxor(Share1,Share2);
I2= bitxor(R2,Sm2);
figure,imshow(I1)
figure,imshow(I2)
```

The above mentioned method is valid for any n number of shares and secrets. We need to carry out the above mentioned progressive XORing for all the secret images. We can see in Figure 4.8 the experimental result of XOR-based multi secret sharing with random shares as shown in codes 4.1 and 4.2. Here Figure 4.8 (a) and (b) are two multitone secret images which are to be secretly shared. Figure 4.8 (c),(d) and (e) are master shares, share 1 and share 2, respectively. Figure 4.8 (f)and (g) are the recovery results, which show the effectiveness of the approach. We see here that we are recovering exactly the same secret images which were shared.

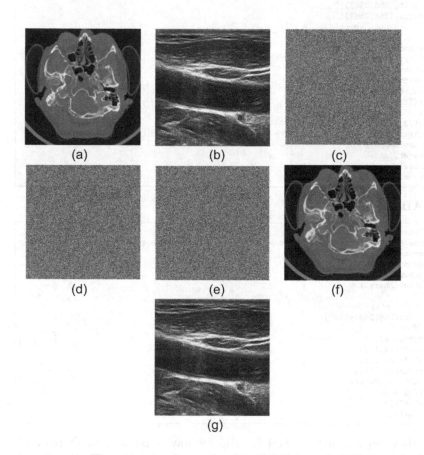

FIGURE 4.8 Experimental result for XOR-based MSS with random shares (*Courtesy of Dr. Luca Saba, Radiology Department, University of Cagliari, Italy*).

4.2.6   XOR-Based MSS with Unexpanded Meaningful Shares

This method can be extended by providing the features of meaningful shares. To do so we need to first generate random shares. These random shares will be converted into a meaningful version. Since we are allowed to do calculations at the receiver end, there may be many methods to do so. Readers may create their own effective approach. For example we can use the same number of multitone cover images and the generated random shares are embedded into cover images as watermarks (as we will see in the next part of the book). Because of the property of a watermark, we cannot see the watermark but the cover images can be seen.

4.2.7   Development of Verifiable Visual Cryptography

Verifiability is an additional feature of shares. Verifiability can be added to all kind of shares whether generated by computationless or computation-based VC. Verifiability is introduced into the shares after their generation. If any share has the property to authenticate itself, then we say that approach is a verifiable VC approach. It means verifiability can be combined with any of the standard visual cryptography approaches to make a hybrid approach. For example, we can make verifiable progressive visual cryptography where shares generated by a progressive VC approach are again processed into the verifiable visual cryptography approach to make them verifiable. Similarly, one can make verifiable multitone visual cryptography, verifiable multi secret sharing approach and so on. If a VC approach is a hybrid (combined with verifiable visual cryptography)approach then before stacking of the shares (for computationless)or processing of the shares (for computation-based), we must check the shares for their authenticity. Verifiable shares are also known as cheating immune shares because we can identify any of the tampered shares. We can also enhance the accuracy of the verifiability so that we can also determine the location of the tampering in the share.

Points worth remembering:
Verifiability is the property of the shares. It can be applied on shares generated by any VC approach.

Points worth remembering:
One can create a hybrid approach by combining verifiability with any VC approach.

4.2.7.1 Providing verifiability into the shares generated by computationless VC approaches

As we know, all shares which are generated by the computationless VC approaches are always binary in nature. A binary image has only two intensities either black or white. These intensities can be represented by a single bit only (1 for black and 0 for white). It means there is no space for embedding of the verifiable bit into the pixel itself. We must replace the share's bit by the verifiable bit. Hence we have to generate a cumulative verifiable bit for a single group of secret pixels. The number of verifiable bits must be decided in such a way that it can properly authenticate the shares, as well as due to secret pixel replacement of the shares by verifiable bits, the quality of the recovered secret must not be degraded. There may be many methods to provide verifiability to the shares. Here we are suggesting one basic method. Readers can create their own efficient method.

4.2.7.2 Steps for adding verifiability into the shares generated by computationless VC approaches

1. Here we assume that before generation of the shares using any of the VC algorithms, a binary secret image has been preprocessed. Thus, the secret image has been converted into the number of blocks of size 2×2 having either complete black or complete white pixels.

2. Let us assume that we are generating 4 out of 4 VC approaches. Therefore, the secret will only be revealed when all four shares are stacked together.

3. Verifiability is the feature of the share hence we have to generate verifiable bits for each share separately. Once the verifiable bits are generated for the shares, then we have to embed them within the respective shares.

4. As we know, shares themselves are binary in nature hence there is no space within the pixels of the shares to embed additional bits,

5. To generate and embed the verifiable bits, first we divide the secret image as well as shares. Suppose we have divided the images into non-overlapping blocks of size 8×8. Each pixel contains only a single bit because it is a binary image. Hence there will be $8 \times 8 = 64$ bits for which we need to generate an authentication or verifiable bit. We can also take smaller or larger block sizes as per our requirement. Each share will have its own set of authentication bits because verifiability is the property of each share.

6. We have to generate a single authentication bit for sixty-four pixels of a single block of each share. Let the size of the shares and secret image be $M \times N$, then the total number of authentication bits is $\frac{M \times N}{64}$ for each share.

7. We need to develop an algorithm for authentication bit generation. This authentication bit must be highly coupled with its generator which is sixty-four bits, so that if any change is made in the sixty-four bits by any method, then this authentication bit must be changed.

8. There may be many ways by which we can generate authentication bit for a block. Let authentication bit for a block be Au. We need to bind it with the sixty four bits of a block denoted by B. Let us consider that we have used the following method to generate Au:

$$Au = (\sum_{i=1}^{8} \sum_{j=1}^{8} (B(i,j))) mod\ 2 \qquad (4.1)$$

In the above mentioned method, we are dealing with the hamming weight of the block. We can also relate the pixels to their neighbours by using the following method.

9. First of all convert the block into a vector V of size 1×64 and then apply the following:

$$Au = (\sum_{i=1}^{63} (V(i) \oplus V(i+1))) mod\ 2 \qquad (4.2)$$

The above mentioned equation also returns a single authentication bit which is coupled with its neighbour. One can make another relation which is generated by the position of the sixty four bits. There are many methods which are nothing but the features of that block and they can be used to generate authentication bit Au.

10. The same process will be applied for each block of each share. Now the problem is where to hide this authentication bit in a block so that it does not affect the quality of shares as well as the recovered secret.

11. In Figure 4.9, we suggested one such method. There may be many more methods. As we know, a secret image will always be preprocessed before share generation. Now we have to create an indexing of all completely black blocks of size 2×2 in a block of size 8×8 of each share.

12. Secretly choose an index of an all black block from a preprocessed secret image. Now the same position will be tracked in each share for authentication bit embedding.

13. Our objective is to embed the authentication bit in such a way that when we stack all the shares, then we get an unaltered recovered block. Since we are embedding the authentication bit on the site of an all black block, hence at the time of stacking, we must get an all black block.

14. To achieve this objective we can use an additional black pixel with the authentication bit. Authentication bit may be either 0 or 1 but an additional black pixel will always be 1. This additional black pixel is positioned in each share in such a way that when we superimpose all the shares, then changes that occurred due to authentication bit embedding must be suppressed. We will get an all black block as it was in the original secret image. The same procedure can be seen in Figure 4.9.

15. In order to verify the authenticity of a share, we must recalculate all the authentication bits of the blocks of the share. These authentication bits will be compared with the corresponding extracted authentication bits from the block. If there is a mismatch in any authentication bit of any block, then that share will be declared tampered or corrupted one and the receiver may ask the sender to send the share again.

16. One can track the location of the tampering also. If any recalculated authentication bit is mismatched with its corresponding extracted authentication bit, then the whole block will be treated as a tampered block. In this way, we can track the tampering location also.

Points worth remembering:
To provide verifiability to binary shares, we need to divide the shares into blocks.

Points worth remembering:
Authenticity of shares is always confirmed before the stacking of the shares.

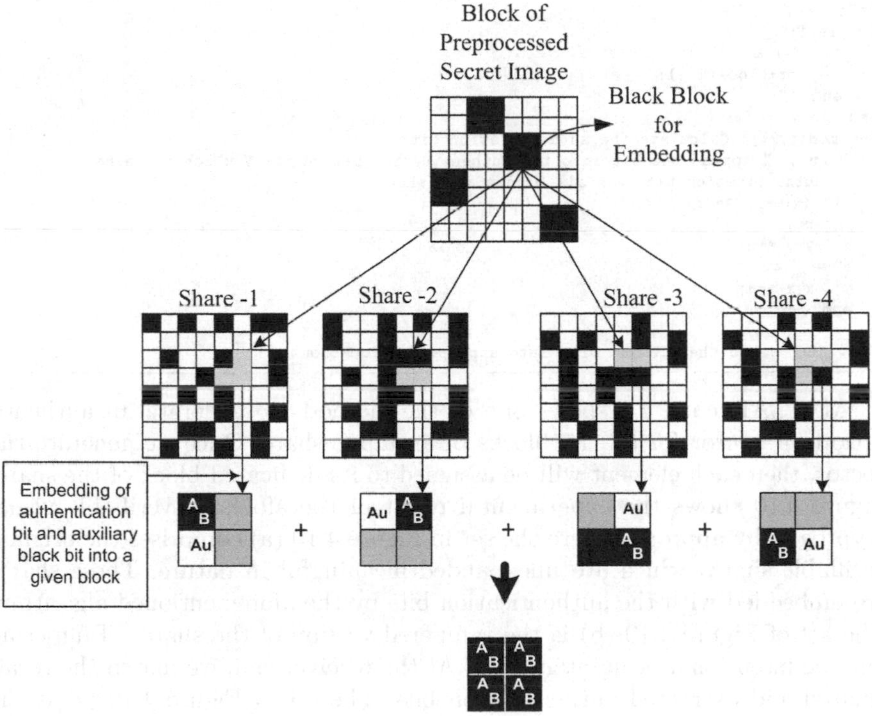

FIGURE 4.9 Authentication bit embedding for computationless VC.

MATLAB Code 4.5

```
%MATLAB Code 4.5
%MATLAB code for authentication bit generation for each block of the share.
Share=double(imread('lena_share.bmp'));
[r c]=size(Share);
m=8;
n=8;
NofoBlock= (r*c)/(m*n); % Calculate the number of blocks
m=8;% mxm is the size of the block
x=1;
y=1;
V=[];
for (kk = 1:NoofBlock)
block=Share(y:y+(m-1),x:x+(m-1));% Assigning the block of size mxm to variable block
 t=0;
 for i= 1:8
     for j=1:8
         t=t+block(i,j);
     end
 end
 Au= mod(t,2);% Calculate the authentication bit
 V=[V Au]; % Append the bit into the authentication bit vector V which contains
     authentication bits for all blocks of a share
     if (x+m*n) >= c
        x=1;
        y=y+m*n;
    else
        x=x+m*n;
    end
end
% Now just embed the element of V into appropriate locations.
```

MATLAB code 4.5 shows one of the methods to generate an authentication bit vector for all the blocks of an input share. Once we generate the vector, then each element will be assigned to its dedicated block of the share. Figure 4.10 shows the experimental result of the aforesaid verifiable visual cryptography approach. Here the set in Figure 4.10 (a) comprises the original verifiable shares which are unexpanded meaningful in nature. These shares are embedded with the authentication bits by the aforementioned algorithm. The set of Figure 4.10 (b) is the tampered version of the shares. Tampering may be intentional or unintentional. At the receiver end, we match the recalculated and extracted authentication bits. The set of Figure 4.10 (c) is the tamper detection results. Here all of the white portion shows the mismatched location between the recalculated and extracted bits of the block. All of the black portion shows the unaltered region. The whole block will be treated as altered if it contains mismatched authentication bits.

4.2.7.3 Adding verifiability into the shares generated by computation-based VC approaches

In the last section we saw how verifiability can be generated into the shares which are obtained by the computationless VC approaches. Now in this section, we fulfil the same objective for the shares generated by computation-based VC. As we know, in this case we have freedom to do any type of arithmetic and logical operations among the shares, hence shares are always mul-

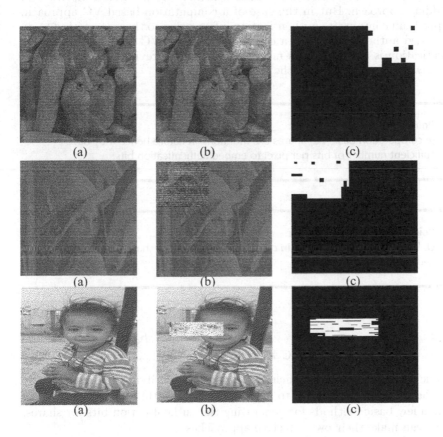

FIGURE 4.10 Example of tampered share and tampering detection results.

titone in nature. Unlike binary shares, the intensity of a multitone image is always represented by more than one bit. Most of the time when we deal with gray scale shares, each pixel of the share is represented by eight bits whereas when we deal with color shares then each pixel of the share is represented by twenty-four bits (eight for each color plane). It means, once we generate the authentication bits, then we have many options to hide or embed that authentication bit into the shares. In computationless VC, each pixel was a single bit so we did not have embedding space which is why we used a block-based embedding approach. But in the case of a computation-based VC approach, each pixel can contains its authentication information so there is no need to do block-based authentication. In a computation-based VC approach, pixelwise authentication is done which is more accurate than previous authentications because we can exactly locate the tampered pixels.

Points worth remembering:
In computation-base VC, shares are multitone in nature hence they have a sufficient number of bits per pixel to embed authentication bits.

Points worth remembering:
Unlike computationless VC, in computation-based VC, shares can be verified pixelwise.

4.2.7.4 Steps for providing verifiability in the shares generated by computation-based VC approaches

Here one can create a more efficient approach for verifying the authenticity of the shares as we are allowed to do computation. In this section we are presenting a few basic methods for generating the authentication bits for shares. Readers can make their own effective approaches.

1. First, we need to understand the basic relationship between the imperceptibility and LSBs of the image. Actually, we will see all these in detail in the next "Fragile Watermarking" chapter. Here we will provide a little working knowledge so that we can better understand the concept of verifiability.

2. Let us assume that we are dealing with gray scale shares. It means the pixel of each share can be represented by eight bits. Actually, lower bits (LSBs) of the pixels hold less visual information of the image whereas higher bits (MSBs) keep the structural information of the image. It

means whenever we generate authentication bits, we need to embed or hide them in the LSBs of the corresponding pixels of the shares.

3. Imperceptibility is the measurement of similarity. It means when we embed the authentication bits inside the shares, then the visual quality of the shares must not be changed and it must look like the original share. Suppose we are generating a single authentication bit for each pixel of the share. When we embed it into the first LSB of the corresponding pixel, then it can change the original intensity of the pixel by a single value. For example, if the original intensity of a pixel is 50 (binary 00110010), by our method its generated authentication bit is 1. Then after embedding of the authentication bit into the first LSB of the intensity, the modified value will be 51 (binary 00110011). Our human visual system cannot interpret the visual difference between 50 and 51. Similarly, if we dedicate two LSBs of a pixel for embedding, then a pixel can be changed such that by at max the difference of four(00(0) can be 11(3)).

4. Hence first of all we need to decide the number of LSBs per pixel which will act as the embedder for the authentication bit. One more thing to know is that when we increase the number of LSBs for embedding, the imperceptibility will be decreased.

5. When a share is generated and later on it is transformed into the verifiable share, then we need to decide the number of authentication bits and their embedding locations. In this way, the secret recovery at the receiver end will not be affected.

6. An authentication bit is nothing but a bit which is tightly coupled with its originating pixel's value, position and other various features. Thus, if a pixel is changed, then the corresponding authentication bit will no longer valid. Here we are going to discuss a few basic approaches. Readers can create their own efficient approach.

7. Let $I = \begin{bmatrix} 108 & 51 \\ 36 & 15 \end{bmatrix}$ be a multitone share of size 2×2. Now we need to make it verifiable in nature. First of all convert it into its binary form i.e. $I = \begin{bmatrix} 01101100 & 00110011 \\ 00100100 & 00001111 \end{bmatrix}$. Suppose we have decided to embed the authentication bit into the first LSB of each pixel. Now we can make the authentication bit as follows:

$$Au = (\sum_{i=2}^{7}(P_b(i) \oplus P_b(i+1)))mod\ 2 \qquad (4.3)$$

Here $P_b(i)$ is the ith bit of the pixel P, where $i = 1 - 8$. By this method, we are just binding the pixel with its bit sequence. In this equation, the calculation is started from the second bit because the first bit is dedicated to authentication bit embedding. If we include the first bit for authentication bit calculation then we will get an undesirable result at the time of tamper detection. This method for authentication bit calculation will be applied for all pixels of the image. If the size of the image is $M \times N$ then there will be $M \times N$ authentication bits. One bit is dedicated to a single pixel.

8. There are many other methods by which an authentication bit can be generated. For example, we can bind the pixel by its position and generate the authentication bit. If the corresponding rows and columns for a pixel of a given block are $\begin{bmatrix} & 37 & 38 \\ 102 & 108 & 51 \\ 103 & 36 & 15 \end{bmatrix}$.

$$A_{s_1} = ExOr(b^u(Row), b^u(P)) \tag{4.4}$$

$$A_{s_2} = ExOr(b^u(Column), b^u(P)) \tag{4.5}$$

where $u = 2, 1, ..8$ After associating the pixel P with its row and column value we get two vectors A_{s_1} and A_{s_2} of size 1×7. Now we can calculate the authentication bit Au in the following way:

$$A_u = \{\sum_{u=1}^{7}(A_{s_1}^u \wedge A_{s_2}^u)\} mod2 \tag{4.6}$$

9. This calculated Au is tightly bounded with its position and its value. In the future if there is any change in the pixel due to an attack then this Au bit has the maximum probability of being changed.

10. The same procedure will be applied on all the pixels of the image. We can generate more than one authentication bit for a each pixel for better tampering detection rate, but it can reduce the imperceptibility.

11. At the receiver end, we again recalculate the authentication bit for all the pixels as well as extract the embedded authentication bit from the first LSBs of the pixels. Now we compare the corresponding recalculated and extracted bits. If there is any mismatch then that pixel will be marked as a tampered pixel.

One can understand the the process of authentication bit generation for multitone image from Figure 4.11. Here we can see that seven bits (excluding the first LSB) are taken for the authentication bit generation. We can apply any method which can generate an authentication bit which is highly coupled with the corresponding pixel.

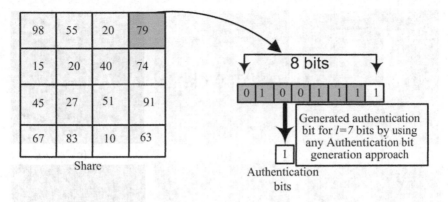

FIGURE 4.11 Authentication bit generation in a mutitone image.

Points worth remembering:
Authentication bits are highly coupled with the pixel value, position or other features of the corresponding pixels.

Points worth remembering:
One or more than one authentication bit can be generated for a single pixel for better verifiability.

Figure 4.12 is the experimental result for the above mentioned method. Here the set in Figure 4.12 (a) is the verifiable multitone secret images. The Set in Figure 4.12 (b) is the tampered version of set (a). Figures in set (c) are the detection results. Here the white pixel shows the tampered location whereas the black pixels are unaltered. One can see that most of the altered pixels are shown as unaltered in Figure 4.12(c) because of lower accuracy of the authentication bit. The development of an efficient algorithm for providing verifiability of shares is an area still under research.

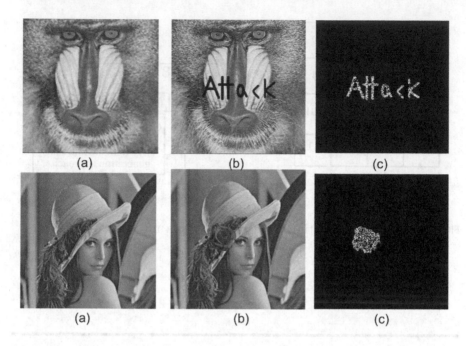

(a)   (b)   (c)

(a)   (b)   (c)

FIGURE 4.12 Experimental result for tamper detection for the shares generated by computation-based VC approach.

SUMMARY

- A computation-based VC approach eliminates a number of *unnecessary encryption constraints* in VC like contrast loss of the recovered secret at the receiver end, binary image as the secret, limitations of share and secret image, etc.

- *Share alignment* is the biggest problem in computationless VC approaches.

- There is a *trade-off between share alignment problems and pixel expansion*. When we increase the pixel expansion, the share alignment problem is reduced and vice versa.

- Computationless VC is based on a logical OR operation whereas most of the time computation-based VC is based on a XOR logical operation.

- In XOR-based VC, the shares are by default unexpanded.

- The size of the basis matrix in XOR-based VC is $t \times n$ where $t = \lceil \frac{n}{2} \rceil$.

- One can share n number of multitone secrets using n number of shares using XOR-based MSS.

- *Reversibility* is a very important property of the XOR operation which is used in multi secret sharing.

- *Verifiability* is a property of the shares. It can be applied on shares generated by any VC approach.

- One can make a *hybrid approach* by combining verifiability with any VC approach.

- In computation-based VC, shares are multitone in nature hence they have sufficient number of bits per pixel to embed *authentication bits*.

- Unlike computationless VC, in computation-based VC, shares can be verified *pixelwise*.

- Authentication bits are *highly coupled* with the pixel value, position or other features of the corresponding pixels.

- One or more than one authentication bit can be generated for a single pixel for better verifiability.

II

Digital Image Watermarking

CHAPTER 5

Digital Image Watermarking: Introduction

CONTENTS

I N this chapter we will introduce and study in detail the different evaluation parameters, properties and various applications of watermarking. We can define watermarking as the practice of imperceptibly embedding a message in that work. Here, two terms are used, one is message and the other one is work. The context of both terms will be discussed in detail in the upcoming

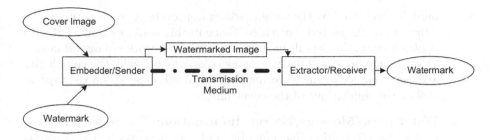

FIGURE 5.1 Basic model for watermarking approach.

sections. There are number of components in the watermarking system and each component plays an important role. Watermarking is becoming more and more popular security approach as it already has a number of applications in diverse areas. Each application can be treated as motivation to make further advancements in watermarking. All watermarking approaches must have some essential properties. Based on these essential properties, we can verify whether a given watermarking approach is valid or not. There are various evaluation parameters to compare watermarking approaches and determine the superiority of one over other.

Hence in this chapter we deal with various theoretical aspects of watermarking which are necessary to understand before developing any new watermarking approach. The **Chapter Learning Outcomes(CLO)** of this chapter are given below. After reading this chapter, readers will be able to:

1. Understand the basics of watermarking.

2. Understand different applications of watermarking.

3. Understand various properties of watermarking.

4. Identify different categories of watermarking based on features.

5. Evaluate different measures of watermarking.

5.1 INTRODUCTION

Watermarking is a technique for embedding secret information (the watermark) into a digital signal (cover) which can later be extracted or detected to verify the authenticity or identity of the owner. Before studying anything about watermarking, we must learn all the basic components and terminology related to watermarking which we will encounter throughout the book. Figure 5.1 shows the basic model of a watermarking approach. Various components used in this model are as follows:

1. **Cover/Host Work:** Cover or host works are those digital signals which

must be protected by the watermarking approach. A cover work may be either an image or audio or video. Since in this book, we only deal with digital images, the default meaning of cover/host work is a digital image. Cover image, may be binary, gray or color which is why the complexity and performance of the watermarking approach depends on the type as well as the dimensions of the cover image.

2. **Watermark/Message/Secret Information:** The watermark/message is the information that must be hidden in the cover image. A watermark is used to protect the cover image and it may be related/unrelated to the cover image in which it is going to be hidden. The relationship of the watermark to the cover work depends on the type and application of the watermarking approach. A watermark may be derived from the cover image or may be some other meaningful image. It may also be a random image or random pattern of bits.

3. **Embedding System/Sender:** The sender or embedding system is the one which embeds the watermark into the cover image. All proposed watermarking approaches are executed by this (sender/embedder) entity only. The execution time of the embedding system depends upon the watermark embedding algorithm and the nature of the cover and watermark.

4. **Watermarked Image:** Watermarked images are nothing but the cover images embedded with the watermark.

5. **Transmission Channel:** This is the medium by which the watermarked image travels up to the receiver end.

6. **Extraction System/Receiver:** This is the entity where the watermark extraction algorithm is executed to extract the watermark and thus obtain the original image.

The aforementioned entities are the basic components of any watermarking scheme. Each component is related to others directly or indirectly. The watermark and cover image both are separate entities but both may be interrelated as per the type of watermarking. The watermark and cover image are both treated as input to the embedding system or sender. The embedding system executes the watermark embedding algorithm and outputs the watermarked image. This watermarked image is transmitted via any secure transmission medium to the receiver. The extraction system separates the watermark from the watermarked image.

Points worth remembering:
The cover image, watermark, sender and receiver are the main components of a watermarking approach.

5.1.1 Significance of the word "Watermark"

The name "Watermark" itself seems amazing and holds some meaning and significance. Let us consider a paper where some text is written. Now if we put some drops of water on the paper, then we can still see the contents of the paper through the water drops. We can say that the contents are covered with the transparent protection of the water. The same thing happens in the case of hidden or invisible watermark. We embed the watermark into the content of the cover image but we can still see the content of the watermarked image clearly. The embedded watermark plays the role of protector of the cover image. That is why this type of mark is called a watermark as it resembles the behaviour of water in the above example.

Points worth remembering:
A watermark means a drop of water on written text or image on the paper.

5.1.2 Importance of Watermarking

Digitization of multimedia content is increasing day by day across the Internet. Tampering of digital content is becoming easier in practice. To protect the integrity of digital content and also its ownership assertion, the concept of watermarking came into the picture. The Internet is inexpensive, instantaneous and an excellent distribution system of digital media like images, audio and video. There is a high risk of man in the middle attacks on the content in digital media. These attacks are intended for either ownership spoofing or for altering the original content of the digital media. Most of the owners of digital content suffer from the problem of piracy of digital content, because these can be easily duplicated without the consent of the owner. Digital watermarking is the best solution for the problem of copyright infringement of digital content. Digital watermarking schemes assure the ownership assertion and integrity of the digital content with much less complexity and cost. It can also provide many more features such as content recoverability with a little bit of computation. Cryptography was the first approach that was used to protect digital content. In cryptography, digital content is encrypted before delivery, and a decryption key (which is most of the time a symmetric key) is provided to legitimate customers of that content. Then this encrypted version of the digital content can easily be transmitted over the Internet. As the contents are encrypted, it means no one can see the content (confidentiality is achieved). Here our assumption is that the customer is honest and we blindly trust that he will not redistribute the content illegally after decryption. Actually, by using cryptography, we cannot monitor the content as well as customer after selling the content. This is the normal modus operandi of the pirates where they actually purchase the first copy of the content legally, and after that they redistribute the decrypted version of the content. Thus, when we protect the content using cryptography, it will no longer be secure after decryption.

Digital watermarking is a strong substitute for the cryptography for the protection of digital content. It modifies the content in such a way that even after many normal modifications like rotation. cropping, denoising, compression etc of the digital content, we can still make an ownership assertion. Watermarking techniques can easily be implemented on hardware so that digital content can be protected at run time (during recording). Cryptography in addition to watermarking can further enhance the power of the watermarking technique.

5.2 WATERMARKING APPLICATIONS

There are a number of areas where watermarking can be used to provide or enhance security. An efficient watermark has various properties and each property is responsible for an application. Some applications are noted below.

5.2.1 Proof of Ownership

By using a watermarking approach, one can prove ownership of any digital image. The sender must register the image with the copyright office or any trusted third party. The copyright office archives the image, together with information about the rightful owner. When a dispute arises, the copyright office can be contacted to obtain proof of rightful ownership.

5.2.2 Ownership Identification

In the aforementioned application, we use the intervention of the copyright office to prove ownership. Ownership can also be verified without intervention of the copyright office. An additional watermark can be used for ownership assertion. It provides complementary copyright marking functionality as it becomes an integral part of the content. The identity of the copy right owner is embedded as the watermark. At the time of any conflict, that watermark is extracted to confirm ownership. Figure 5.2 shows the copyright information embedded into the cover image.

5.2.3 Broadcast Monitoring

A very traditional and low-cost method for observing the broadcast signal is monitoring by a human observer but this is a much less efficient and costly method. That is why there must be some alternative method which is fully automatic. This technique may be categorized in two ways: one is active monitoring and the other one is passive monitoring. Passive monitoring systems directly recognize the content which is being broadcast whereas active moni-

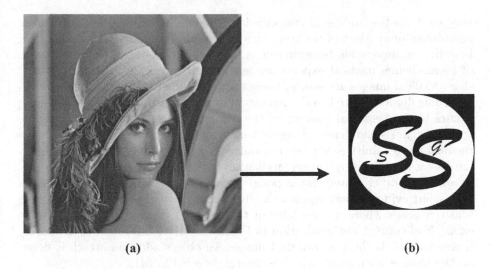

(a)　　　　　　　　　　　　　　(b)

FIGURE 5.2 Copyright information embedded into the cover image: (a) Cover image, (b) Watermark (copyright information).

toring systems rely on the associated information that is broadcast along with the content. A passive system directly compares the signal with the already saved signals. When the comparison locates a match, then the broadcast signal can be identified. This is a very simple and effective method for automated broadcast monitoring. In this case we do not need any associated information in the broadcast. Actually, this method does not require any cooperation with the advertisers. But there are many potential problems with implementing passive monitoring systems. First of all a comparison of a broadcast signal with the stored one is not an easy task. In practice, signals are divided into smaller units, such as individual frames of video, and searching is carried out for them in the database. As we know that a single frame of video may contain a million bits of information, which is why searching of bitwise patterns is impractical. We can say that a passive monitoring system cannot search for an exact match in its database hence designing a passive monitoring system that is 100% reliable is not possible. In the case of an active system, identification information is also placed with the broadcast signal. This identification information is placed in the signal in such a way that it does not affect the quality of the signal. This identification information is nothing but the watermark and just by only matching the watermarks one can compare the two signals.

5.2.4   Content Authentication

A watermark can be used to verify the authenticity of a given cover image. In this application we can check the integrity of any watermarked image to see if it has been tampered or not. A digital cover image can be tampered with very easily in the middle of transmission. The alteration may be intentional or unintentional. Most of the time, this type of alteration must be tracked for healthy communication between sender and receiver. For example, in the case of telemedicine, medical experts are not present physically at the diagnosis sites. Medical images are sent to them via the Internet for their expert advice regarding disease related to the patient. If during transmission, a little change occurs in the principal content of the medical image, then it may cause a wrong diagnosis and hence wrong treatment for the patient. That is why these medical images must either be transmitted via secure channels or they must have some integrity verification mechanism so that experts can distinguish between the tampered and untampered medical images. Similarly, tampering with court evidence may result in a wrong conviction. There are many other sensitive areas where authenticity of the image is essential. We can see an example of content authentication in Figure 5.3. Here image (a) is the cover image and (b) is the watermarked image. An object addition attack is done on the image shown in (c) which is clearly detected in (d).

5.2.5 Tamper Recovery

Sometimes, due to intentional or unintentional attacks, some principal contents of the image are removed and they must be recovered for proper understanding of the image at the receiver end. Using watermarking, we can make an image capable of recovering itself at the receiver end even if some tampering is carried out afterwards. A tamper recovery example is shown in Figure 5.4 where (b) is the tampered image of (a). Figure 5.4 (c) shows the tamper detection result whereas (d) is the recovery result which is achieved through a watermarking approach.

5.2.6 Transaction Tracking

Sometime, an authentic user of digital media may work as an illegal reproducer of that work. The work is generated for that authentic user only. This scenario can be understood by an analogy. Suppose a user has purchased licensed software which is only valid for a single user. Now that user has the whole setup of that licensed software including the license key. It is at his discretion whether he illegally redistributes the software or not. If that user redistributes the software illegally, then in such a scenario, we must have some mechanism by which one can trace the user who is illegally reproducing the work. In this application of watermarking, different watermark is embedded into each replica of the cover work. For example if the work is illegally redistributed or sold by the authentic user, then the owner can be found out easily.

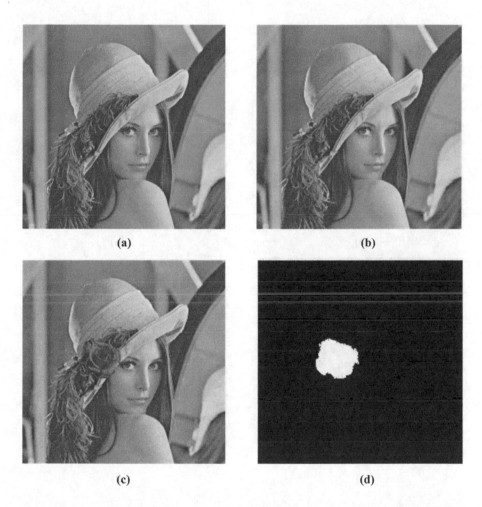

FIGURE 5.3 Example of content authentication using watermarking:(a) Cover image, (b) Watermarked image, (c) Attacked image, (d) Result of tamper detection.

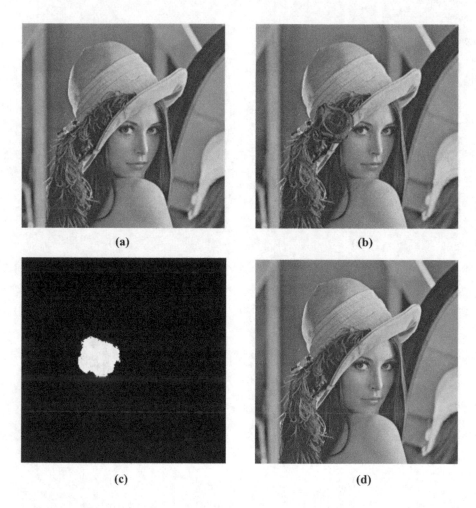

FIGURE 5.4 Example of content authentication and tamper recovery using watermarking: (a) Watermarked image, (b) Tampered version, (c) Tamper detection result, (d) Recovery result.

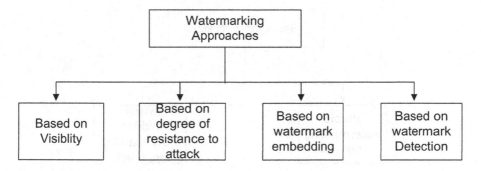

FIGURE 5.5 Classification of watermarking techniques.

5.2.7 Copy Control

In this application, our aim is to prevent people from making illegal copies of copyrighted content. In this case watermarks (never copy bits) are embedded in the content itself, so it will be in every representation of the content. If every recording device would be enabled with a watermark detector, the devices could be made in such a way that they prohibit recording whenever a never-copy watermark is detected at the input. For example, most of the branded CD/DVD writers are enabled with this facility in order to prevent misuse of their CD/DVD writing features. They have capability of reading never-copy watermark bit which are inserted into each copyrighted material. When we try to make a duplicate copy of the copyrighted CD/DVD using a watermark detector CD/DVD writer, it will read its never-copy watermark bit and stop copying.

5.2.8 Device Control

We can enhance the power of a copy control application which refers to device control. There are many applications where devices react according to detected watermarks in the content. This is somewhat different from the copy control application as it adds more information to the content rather than preventing copying. A very latest example of device control is the Digimarcs Mobile system which embeds a unique identifier into printed images that are going to be distributed such as advertisements in magazine, tickets, and so on. When these images are recaptured by camera on mobile phones, the identifier which is embedded into the images is read by the pre-installed software on the phone. That identifier will redirect the user to the associated website with that image.

5.3 CLASSIFICATION OF WATERMARKING TECHNIQUES

Based on the various features, watermarking approaches can be classified as shown in Figure 5.5 and noted in the following list.

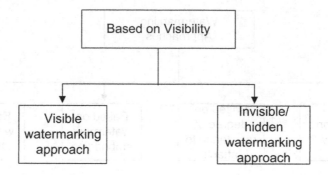

FIGURE 5.6   Classification of watermarking techniques based on visibility.

1. Based on visibility

2. Based on degree of resistance to attacks

3. Based on watermark embedding

4. Based on watermark detection

5.3.1   Based on Visibility

When we embed a watermark into the cover image, the watermark may be either visible or invisible on that image. Hence watermarking is categorized as shown in Figure 5.6.

5.3.1.1   Visible watermarking

Those watermarking approaches in which the embedded watermark is visible on the watermarked image are called visible watermarking approaches. There are many areas in which we need such types of visible watermarks so that just by seeing the watermarked image one can identify the watermark. Suppose we have some confidential papers or images and we want to ensure that if anybody makes duplicates, the ownership information remains intact on the original document. In that case, we embed a visible watermark on the document which actually overlaps with the original content of the cover and the intensity of the watermark is very similar to the content. If anybody removes the watermark, then he will definitely loose some original content of the cover also. Figure 5.7 shows an example of a visible watermark, and Figure 5.7 (b) shows Indian currency in 100 rupees where the note contains a visible watermark of Mahatma Gandhi. This is a special kind of watermark which appears only when we apply light to the embedded portion.

(a)   (b)

FIGURE 5.7 Example of visible watermark.

5.3.1.2 Invisible/hidden watermarking

Unlike the visible watermarking approaches, invisible or hidden watermarking approaches are those in which the embedded watermark does not appear on the watermarked image. We can only extract the watermark using an extraction mechanism. These types of watermarks are used where we do not want to compromise the imperceptibility of the watermarked image.

Points worth remembering:
Based on visibility, a watermark can be categorized into two types: visible and invisible watermarks.

5.3.2 Based on Degree of Resistance to Attacks

Watermarks can also be classified based on their degree of resistance to attacks. Some watermarks cannot tolerate an attack and some can tolerate very efficiently. We classify the resistance in three ways as shown in Figure 5.8.

5.3.2.1 Robust watermark

In the case of a robust watermark, a watermark must survive in an environment where there are a number of intentional or unintentional attacks. Robustness is essential in order to check the ownership of the given cover image. The watermark is embedded into the cover image in such a way that if the content of the watermarked image is tampered with in any way, the associated robust watermark will remain as is. At the receiver end, the existence of the watermark confirms the ownership of the associated cover image.

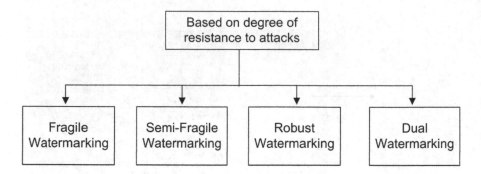

FIGURE 5.8 Classification of watermarking techniques based on degree of resistance to attacks.

Hence our requirement is to make a type of watermark that is highly resistant to intentional or unintentional attacks. All those approaches which generate a robust watermark are called robust watermarking approaches.

5.3.2.2 Fragile watermark

A fragile watermark is one which will be destroyed as soon as any modification is done to the watermarked image. In case of robust watermarking, fragility is highly undesirable because the robust watermark must survive in environment of many intentional or unintentional attacks. Fragility is essential in order to check the integrity of the given cover image. A fragile watermark is embedded into the cover image in such a way that if the content of the watermarked image is tampered, the associated fragile watermark with that content is also tampered or destroyed. At the receiver end, the unavailability of the watermark confirms that the associated principal content of the cover image has been tampered. All those approaches which generate fragile watermark are called fragile watermarking approaches.

5.3.2.3 Semi-fragile watermark

On the basis of the tolerance level to attacks, fragile watermarks can be further classified into another category called a semi-fragile watermark. Semi-fragile watermarks are those watermarks, that are robust watermarks for unintentional attacks and fragile for intentional attacks. A threshold is used to separate the amount of alteration of content. If alteration of content is less than the threshold value then, it will be treated as an unintentional attack, otherwise it is treated as an intentional attack. Semi-fragile watermarks change their role according to the amount of alteration. They are a combination of robust and fragile watermarks.

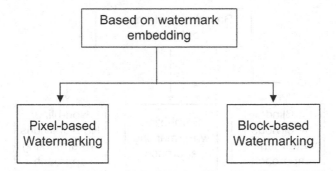

FIGURE 5.9 Classification of watermarking techniques based on watermark embedding.

5.3.2.4 Dual watermarking

A dual watermarking is an approach which has functionality as both types, namely, fragile as well as robust. One should not be confused between semi-fragile watermarking and dual watermarking schemes. They are entirely different. Semi-fragile watermark also has properties of both the robust and fragile watermark but at a given time it acts as either robust or fragile. Dual watermarking is totally different because here we can enjoy the properties of both the robust and fragile watermarks simultaneously.

Points worth remembering:
Based on the degree of resistance to attacks, a watermark can be categorized into four: fragile, semi-fragile, robust and dual watermarks.

5.3.3 Based on Watermark Embedding

A watermarking approach can also be classified based on the the the way the watermark is embedded. It can be categorized in two ways as shown in Figure 5.9.

5.3.3.1 Block-based watermarking

As its name suggests, in this category, a watermark is generated for individual blocks of the given cover image. First of all the whole cover image is divided into non-overlapping blocks of fixed size. After that, the watermark is embedded into each block. The main advantage of this type of watermarking scheme is its low complexity because we do not need to process each and every pixel of the cover image.

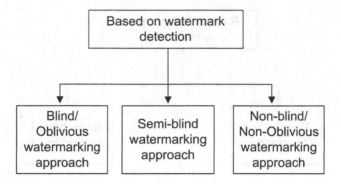

FIGURE 5.10   Classification of watermarking techniques based on watermark detection/extraction.

5.3.3.2   Pixel-based watermarking

In this category of watermarking schemes, a watermark is embedded into each pixel of the cover image. The watermark may be of the same size as the cover image. The complexity of this scheme is high compared to block-based watermarking approaches. Both pixel-based and block-based approaches have their different roles for providing ownership verification and authentication.

Points worth remembering:
Based on embedding, watermarks can be categorized into two ways: pixel-based and Block-based watermark.

5.3.4   Based on Watermark Detection/Extraction

This phase is executed on the receiver side. At the receiver end, we extract/detect the watermark. Based on the method of detection of the watermark, it can further be classified into three ways as shown in Figure 5.10.

5.3.4.1   Non-blind/Non-oblivious watermarking

In this category of watermarking approach, we require the original cover image for extraction or detection of the watermark at the receiver end. This class of watermark is highly undesirable because it increases the overhead for storing and transmitting the cover image along with the watermarked image.

5.3.4.2 Semi-blind watermarking

In this category of watermarking, instead of the whole cover image, we require only some side and abstract information of the cover image at the receiver end, to detect or extract the watermark. Sometimes we embed the watermark in some region-of-interest (ROI) of the cover image (see in detail in the next chapter). In such cases we need to keep track of the embedding locations of the watermark in the cover image. This track record can be treated as side or abstract information of the cover image and it is required at the receiver end in order to detect or extract the watermark. We need to transmit this side information through a secure channel or this information must be encrypted to prevent it from any middle attack.

5.3.4.3 Blind/oblivious watermarking

This category of watermarking is desirable because it does not require any information related to the cover image at the receiver end during the watermark detection or extraction process. Watermark embedding and extraction algorithms are well known to all, hence to prevent the unauthorized extraction or embedding of invalid watermarks into a cover image we need some symmetric key known by only the sender and receiver. So in case of blind watermarking, we need only symmetric key at the receiver end which has been used during embedding of the watermark.

Points worth remembering:
Based on detection, a watermark can be categorized into three ways: blind, semi-blind and non-blind watermark.

5.4 PROPERTIES OF WATERMARKS

There are six properties in ideal watermarking as shown in Figure 5.11 which cannot be ignored during the development of a watermarking approach. These properties may vary according to the types of the watermark. For example if we need watermarking for copyright protection, then the fragility property is invalid. Similarly, if we talk about fragile watermarking the robustness property is invalid for that. There are various other characteristics of watermarking approaches which we will see in more detail.

5.4.1 Robustness

This property is dedicated for the robust watermarking scheme. Robustness is required because it has the ability to resist against various type of intentional or unintentional attacks. There are many routine types of image processing

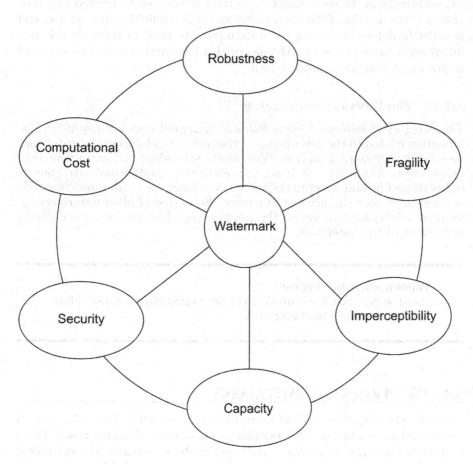

FIGURE 5.11 Various properties of watermarking.

like noise removal, compression, rotation, resizing etc which are sometimes desired for a particular application. Hence our robust watermark must be sustainable in spite of having these image processing operations. Actually, sometimes these image processing operations are made not for intentional removal of the watermark but for enhancement of the quality of the watermarked image. That is why our robust watermarking approach must tolerate all these unintentional attacks as well as a satisfactory number of intentional attacks.

5.4.2 Fragility

This property is dedicated for fragile watermarks. A fragile watermark plays the role of a hash function for images (see more detail in the next chapter). Like a hash function, the generated watermark must be strongly related to the principal content of the cover image so that if any tampering is done on the watermarked image, it must be reflected by its fragile watermark. In this way, we can simply mark the tampered and untampered pixels of the given image. Mapping of fragile watermark with the principal content of the given cover image depends upon the fragility or sensitivity of that watermark for various intentional or unintentional attacks. High fragility tends towards a good fragile watermarking approach. An ideal fragile watermark is consistently sensitive enough for even a single bit alteration of any pixel in the watermarked image.

5.4.3 Imperceptibility

The similarity between an original cover image and its watermarked image is called imperceptibility. Our objective is to design a watermarking scheme that has high imperceptibility between the cover and watermarked image. Imperceptibility can be measured by various parameters. These parameters may be subjective or objective.

5.4.4 Capacity

This property can be considered for all types of watermarking. Capacity is related to the amount of watermark data which can be embedded into the cover image. Ideally, the capacity must be as high as possible so that we can embed more information related to watermark. But there is a trade-off between capacity and imperceptibility. When we increase the capacity of watermark embedding then it will reduce the imperceptibility and also the quality of the watermarked image. Similarly when we reduce the capacity, then imperceptibility of the watermarked image with respect to the cover image will be high.

5.4.5 Security

A given watermarking approach must be secure enough for any kind of intentional or unintentional attack. We need to make a type of watermarking approach in which cryptanalysis is not possible. By this way we can protect the watermark by unauthorized addition or removal of the watermark.

5.4.6 Computational Cost

The computational cost of any watermark embedding or extraction approach should be as low as possible. There is a trade-off between computational cost and accuracy of the watermarking approach. If we reduce the cost of the watermarking approach as in a block-based watermarking scheme, then we compromise the accuracy. Similarly if we increase the computational cost then it definitely will return a good and accurate watermarking scheme. We need to develop an approach that is balanced.

Points worth remembering:
Six essential properties of a watermark are robustness, fragility, imperceptibility, security, capacity and computational cost.

5.5 ATTACKS

Attacks are nothing but modifications of the pixel intensities. Sometimes these modifications are necessary for better visualization/storage of images such as image enhancement/compression, and sometimes pixel intensities are changed unknowingly. There are cases where modifications are done intentionally such as cropping and removal or addition of an object in an image. This is why we can consider these operations an attack.

5.5.1 Types of Attacks

Attacks may be broadly classified into two types viz intentional and unintentional attacks as shown in Figure 5.12.

5.5.1.1 Intentional attack

In this category, an image is tampered with intentionally at the time of its storage or at the time of transmission which is also known as a man in the middle attack. As we can see in Figure 5.13, a rose is pasted intentionally over the hat of Lena. In this case, the amount of tampering may be very high or very small based on the intention of the attacker. An attacker may capture the image and apply a man in the middle attack on it. Once the

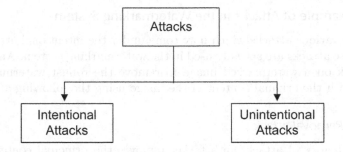

FIGURE 5.12 Types of attacks.

(a)   (b)

FIGURE 5.13 Example of intentional addition of an object.

image is intentionally altered, then we may be the victim of a wrong ownership assertion or unauthentic image reception at the receiver end. Hence we need to develop a type of watermarking scheme that can detect an intentional attack of any type.

5.5.1.2 Unintentional attack

In this class of attack, an image is tampered during the transmission phase. Most of the time an unintentional attack deals with the insertion of noise due to small signal fluctuations during transmission. As we know that any data, image, audio or video, is transmitted over the network in the form of bit streams. Hence there is a high probability of toggling the bits of some data due to any number of obvious reasons. Noises may be Gaussian, Poison, Salt & Pepper etc. Unintentional attacks may also contain usual processing task like zooming, cropping, compression, etc.

5.5.2 Example of Attacks in the Watermarking System

There are various attacks which may come under the intentional attack category. These attacks are actually used in the watermarking system. An attacker can attack on a watermarked image to remove the robust watermark or to tamper with the original content of the image using the following attacks.

5.5.2.1 Removal attack

In this category of attack, an attacker removes the principal content of the image in order to mislead the user. During removal of the content it may be possible that the copyright information of the owner associated with the cover image may get changed.

5.5.2.2 Addition attack

In this attack an attacker may add or write additional objectionable/unobjectionable content on the watermarked image which has nothing to do with watermark information. Figure 5.13 is an example of an object addition attack.

5.5.2.3 Cryptographic attacks

These attacks aim at cracking the security methods in the watermarking schemes and thus finding a way to remove embedded information or to embed misleading watermarks. For example, brute force attacks find secret information through an exhaustive search.

5.5.2.4 Copy paste attack

In this category of attack an attacker may duplicate the watermarked data illegally using a copy and paste operation. Thus a duplicate copy of the watermarked image must also contain the watermark information as embedded in the original one.

5.5.2.5 Print scan attack

This category of attack is very challenging. In this case, suppose we have a watermarked image W and we take a printout of it. The printout version is denoted by P. Now we scan P and make a soft copy of the image. Let this scanned version of the image be denoted by S. Now the challenge is that the watermark which was embedded into the W must be present in S. Considerable research is under way on this topic to handle this class of attack.

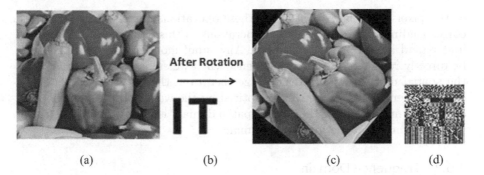

| (a) | (b) | (c) | (d) |

FIGURE 5.14 Example of geometric attack.

5.5.2.6 Geometric attack

This attack causes the loss of synchronization (usually loss of the image coordinates) between the watermark embedded in the cover image and the watermarked image. Usually this attack involves all distortions on the image generated by affine transformations like translation, rotation, scaling, shearing, cropping, line or column removal etc. Figure 5.14 shows an example of a geometric attack where (a) shows the watermarked image embedded with the watermark (b). Then the image is rotated by some angle shown in (c). We can see that the extracted watermark is totally destroyed as shown in (d). That is why a watermark must be robust enough to tolerate such types of geometric attacks.

Points worth remembering:
Attacks are broadly classified into two types: intentional and unintentional attack.

5.6 WATERMARKING DOMAIN

A cover image can be represented in either of the domains viz the spatial domain and the frequency domain. Each domain representation has its own benefits. We need to choose the domain which will be more suitable for a particular type of watermark embedding. First of all we should learn about both domains.

5.6.1 Spatial Domain

In this domain we directly deal with the intensities of the pixel. We can directly apply additive, multiplicative or subtractive operations on intensities of the pixels and the reflections of these operations can be observed in the corresponding images. Pixel-based operations in this domain are very easy to understand as well as to implement. Any small modification on a pixel will be directly reflected on the image. Hence when we embed the watermark into the spatial domain of the cover image then even the smallest modifications on the watermarked image can be noticed by the watermark (see in the next chapter for more detail). Hence the spatial domain is more suitable for fragile watermarking than the frequency domain.

5.6.2 Frequency Domain

In this domain we do not apply direct operations on the pixel intensities. Here we take the transformations (Fourier, DCT, DWT etc) of the images. After getting the transformation coefficients, we apply the operations to them. In the upcoming chapter we will see the concept behind this in more detail. Actually each transformation coefficient of the image is highly dependent on all pixel intensities of the image. Hence we can recover a major portion of the image even after tampering. That is why this domain is used for a robust watermarking approach.

Points worth remembering:
Watermarking approaches may be implemented in two ways either in the spatial domain or the frequency domain.

5.7 MEASURES OF EVALUATION

The performance of watermarking approaches are evaluated with various parameters depending upon the type of watermarking. The first measure for the comparison of two watermarking approaches is based on imperceptibility. Here we have to check that the watermarking approach has good imperceptibility between the watermarked and cover image. To measure the imperceptibility we have various parameters like Peak Signal to Noise Ratio(PSNR), Root Mean Square Error(RMSE), Mean Square Error(MSE), etc. All these parameters deal with the change in pixel intensities of the image. Next if we deal with binary images as copyright (for robust watermarking) then we need some other parameters. The tamper detection rate in case of fragile watermarking can be judged in different ways. So we can say that there are basically two classifications of evaluation parameters, namely, subjective and objective as shown in Figure 5.15 which are used to measure gray/color and binary images, respectively. These parameters are used to judge the proposed watermarking approach.

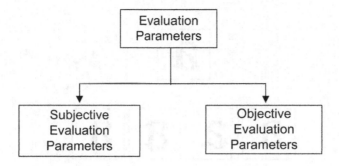

FIGURE 5.15 Classification of evaluation measures.

5.7.1 Subjective Measures

Subjective evaluation parameters like PSNR, SNR, MSE, RMSE etc are well suited for comparing two gray scale or two color images. When our watermark image is of gray scale or color then objective evaluation parameters may give misleading results so subjective evaluation parameters are used in order to compare extracted and original non binary watermark images. The extracted watermark will be more authentic and healthy when its calculated subjective evaluation parameters are near to its ideal values. These parameters are also used to compare the cover and watermarked or watermarked and recovered images.

The average energy distribution caused by embedding watermark on each pixel of the cover image can be calculated as

$$MSE = \frac{1}{mn} \sum_{i=0}^{m-1} \sum_{j=0}^{n-1} (I(i,j) - K(i,j))^2 \qquad (5.1)$$

where MSE is the mean square value which is for $m \times n$ two gray scale images I and K in which one of the images is the original cover image and other one is the watermarked image. The PSNR is defined as

$$PSNR = 10log_{10} \frac{(MAX)^2}{MSE} \qquad (5.2)$$

Here we can see that for two identical images PSNR will be infinite. Hence an efficient watermarking approach will have high PSNR.

Points worth remembering:
Subjective measures are well suited for gray and color scale images.

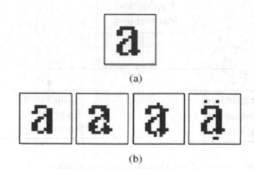

FIGURE 5.16 Distortion in binary image: (a) Original image, (b) Distorted image.

5.7.2 Objective Measures

Sometimes we deal with only binary images. For example in the case of robust watermarking, generally we take binary images as the watermark. At the receiver end, extracted binary watermarks are compared with the original one. This comparison process will always give misleading results when we use Human Visual System (HVS) only. For binary images we need some different parameters which are called objective evaluation parameters. Objective evaluation is performed on the basis of the mutual relations among pixels, pixel values and their locations, hence it is well suited for binary images. While subjective evaluations like Peak Signal to Noise Ratio (PSNR), Signal to Noise Ratio (SNR) and Mean Square Error (MSE) do not match well with subjective assessments. One can understand this through an example shown in Figure 5.16. For instance, a simple binary image is shown in Figure 5.16(a), and four differently distorted images are shown in Figure 5.16(b). Each distorted image has four pixels having opposite binary values compared with their counterpart in Figure 5.16(a). These four distorted images have the same PSNR, SNR and MSE but, the distortion perceived by human visual system (HVS) is quite different.

Let I be the input watermark image and G be the extracted watermark image. N_{TP}, N_{FP}, N_{TN} and N_{FN} are the total number of true positive, false positive, true negative and false negative pixels with respect to I and G respectively. There are many error metrics which are used for objective evaluation of the binary images.

Negative Rate Matrix (NRM): The NRM is based on the pixel wise mismatch between the I and G.

$$NRM = \frac{NR_{fn} + NR_{fp}}{2} \tag{5.3}$$

where

$$NR_{fn} = \frac{N_{FN}}{N_{FN} + N_{TP}} \tag{5.4}$$

$$NR_{fp} = \frac{N_{FP}}{N_{FP} + N_{TN}} \tag{5.5}$$

For an identical pair of images, the value of NRM will be 0.

Recall/Sensitivity:

$$Recall = \frac{N_{TP}}{N_{TP} + N_{FN}} \tag{5.6}$$

For an identical pair of images, the value of recall will be 1.

Precision:

$$Precision = \frac{N_{TP}}{N_{TP} + N_{FP}} \tag{5.7}$$

For an identical pair of images, the value of Precision will be 1.

F-Measure:

$$FM = \frac{2 \times Recall \times Precision}{Recall + Precision} \tag{5.8}$$

For an identical pair of images, the value of F-Measure will be 1.

Specificity:

$$Specificity = \frac{N_{TN}}{N_{TN} + N_{FP}} \tag{5.9}$$

For an identical pair of images, the value of Precision will be 1.

Balanced Classification Rate (BCR)/Area Under the Curve (AUC):

$$BCR = 0.5 \times (Specificity + Sensitivity) \tag{5.10}$$

For identical pair of images the value of BCR/AUC will be 1.

Balanced Error Rate (BER):

$$BER = 100 \times (1 - BCR) \tag{5.11}$$

For an identical pair of images, the value of Precision will be 0.

Structural Similarity Index (SSIM):
SSIM is a Human Visual System (HVS)-based measure. Many different types of HVS models have been developed to measure image quality, however, among all objective measures, the SSIM measure is considered to be the closest to subjective measures.

$$SSIM(x,y) = \frac{(2\mu_x\mu_y + c_1)(2\sigma_{xy} + c_2)}{(\mu_x^2 + \mu_y^2 + c_1)(\sigma_x^2 + \sigma_y^2 + c_1)} \tag{5.12}$$

where μ_x, μ_y, σ_x^2, σ_y^2 and σ_{xy} are the average, variance and covariance for x and y, respectively. The resultant SSIM index is a decimal value between -1 and 1, and the value 1 is only reachable in the case of two identical sets of data.

Distance-Reciprocal Distortion Measure (DRDM):
Let W_m be the weight matrix and i_c and j_c be the center pixel.

$$W_m(i,j) = \begin{cases} 0, & \text{if } i_c = j_c \\ \frac{1}{\sqrt{(i-i_c)^2+(j-j_c)^2}}, & \text{otherwise} \end{cases} \tag{5.13}$$

this matrix is normalized by.

$$W_{Nm}(i,j) = \frac{W_m(i,j)}{\sum_{i=1}^{m}\sum_{j=1}^{m} W_m(i,j)} \tag{5.14}$$

Now

$$DRD_k = \sum_{i,j}[D_k(i,j) \times W_{Nm}(i,j)] \tag{5.15}$$

Where D_k is given by $(B_k(i,j) - g[(x,y)_k])$. Thus DRD_k equals to he weighted sum of the pixels in the block B_k of the original image.

$$DRD = \frac{\sum_{k=1}^{s} DRD_k}{NUBN} \tag{5.16}$$

where NUBN are the non-uniform blocks in $F(x,y)$. For an identical pair of images DRDM will be 0.

Points worth remembering:
Objective evaluation parameters are well suited for binary images.

5.7.3 Other Evaluation Parameters

Besides the aforementioned parameters, some more parameters are also used to judge the efficiency of the fragile watermarking approaches. In the case of fragile watermarking we need to locate the tampered portions or pixels of the watermarked image which cannot be measured by the aforementioned parameters. For this we use two different parameters which are False Acceptance Rate (FAR) and False Rejection Rate (FRR).

5.7.3.1 False acceptance rate (FAR)

Sometimes a fragile watermarking scheme detects altered pixels which have actually not been altered. This scenario is called false acceptance and its rate is called false acceptance rate. For a good fragile watermarking scheme, FAR must be as low as possible.

5.7.3.2 False rejection rate (FRR)

If a fragile watermarking scheme treats an altered pixel as unaltered, then this scenario is called the false rejection and its rate is called false rejection rate. For a good fragile watermarking scheme, FRR must also be as low as possible.

Points worth remembering:
FAR and FRR are useful to measure the effectiveness of fragile watermarking schemes.

5.8   WATERMARKING SCHEME WITH RECOVERY CAPABILITIES

The utility of a fragile watermark may be enhanced if we provide recovery capabilities in it. Till now we have seen that the objective of the fragile watermark is authentication/integrity verification. Once we realize that integrity of a watermarked image is compromised, then it is the general human tendency to know the actual and original principal content of the watermarked image. Hence if we embed some recovery information into the cover image along with the authentication information used for tamper localization, it will be easier to track as well as restore the tampered content. There are two methods to generate the recovery information of the cover image as shown in Figure 5.17.

5.8.1   Recovery Using Spatial Domain

In this category we generate the recovery bits using the spatial domain of the cover image. We know that in the spatial domain, we deal directly with the

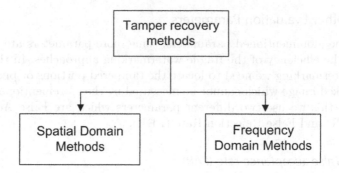

FIGURE 5.17 Classification of recovery approaches.

pixel intensities of the image. Recovery information may either be pixel-based or block-based. Pixel-based recovery gives efficient recovery at the receiver end whereas a block-based approach produces a staircase effect in the recovered image. Spatial domain-based recovery has less complexity when producing recovery bits.

5.8.2   Recovery Using Frequency Domain

In this category, recovery bits are generated using the frequency domain of the image. First of all we transform the image using any transformations such as Discrete Cosine Transformation (DCT), Discrete Wavelet Transformation (DWT) etc. After that, the recovery bits are generated using the frequency coefficients of the image. This type of recovery may also be pixel-based and block-based.

Points worth remembering:
Recovery of a destroyed watermarked image can be done in two ways, either in the spatial domain or the frequency domain.

SUMMARY

- *Watermarking* is a technique for embedding secret information (the watermark) into a digital signal (cover) which can be later extracted or detected to verify the authenticity or identity of its owners.

- Cover image, watermark, sender and receiver are the main components of the watermarking approach.

- A *Watermark* reassembles a drop of water on written text or an image on the paper.

- There are various applications of watermarking including content authentication, ownership verification, tamper recovery, broadcast monitoring etc.

- *Based on visibility*, a watermark can be categorized into two types: visible and invisible watermarks.

- *Based on the degree of resistance to attacks*, watermark can be categorized into four types: fragile, semi-fragile, robust and dual watermarks.

- *Based on embedding*, watermarks can be categorized into two types: pixel-based and block-based watermark.

- *Based on detection*, a watermark can be categorized into three types: blind, semi-blind and non-blind watermarks.

- Six essential properties of the watermark are robustness, fragility, imperceptibility, security, capacity and computational cost.

- Attacks are broadly classified into two types: *intentional and unintentional attacks.*

- Watermarking approaches may be implemented into two ways, either in *the spatial or the frequency domain.*

- *Subjective measures* are well suited for gray and color scale images.

- *Objective measures* are well suited for binary images.

- *FAR and FRR* are useful to measure the effectiveness of fragile watermarking schemes.

- *Recovery* of a destroyed watermarked image can be done in two ways, either in the spatial domain or the frequency domain.

Fragile Watermarking

CONTENTS

I N this chapter, we focus on how a watermark can be used to assist in maintaining and verifying the integrity of its associated cover image. This type of watermark is called a fragile watermark. A digital cover image can be tampered with very easily. The alteration may be intentional or unintentional. Most of the time such types of alteration must be tracked for healthy communication between sender and receiver. For example, in the case of telemedicine, medical experts are not present physically at the diagnosis site. Medical images are sent to them via the Internet for their expert advice regarding disease related to the patient. If during transmission, a small bit change is made to the principal contents of the medical image then this will cause a wrong diagnosis and hence wrong treatment for the patients. For this reason these medical images must either be transmitted via a secure channel or they must have some integrity verification mechanism so that medical experts can distinguish between the tampered and untampered medical images. Similarly tampering with court evidence may result in a wrong conviction. There are many other sensitive areas where the integrity of an image is essential.

Hence in this chapter we deal with various theoretical and practical aspects of fragile watermarks pertaining to the authentication and integrity of the associated cover images. The **Chapter Learning Outcomes(CLO)** for this chapter are given below. After reading this chapter, readers will be able to:

1. Understand the background and need for fragile watermark.

2. Understand the different terminologies related to fragile watermark.

3. Write their own block-based or pixel-based fragile watermarking algorithms.

4. Identify the research gap and research interest in this field.

5. Implement their own scheme in MATLAB.

6.1 INTRODUCTION

A fragile watermark is simply a mark which will be destroyed as soon as any modification is made to the watermarked image. In the case of robust watermarking, fragility is highly undesirable because a robust watermark must survive in an environment of many intentional or unintentional attacks. Fragility is essential in order to check the integrity of the given cover image. A fragile watermark is embedded into the cover image in such a way that if the contents of the watermarked image are tampered with, the associated fragile watermark with that content is also destroyed. At the receiver end, the unavailability of the watermark confirms that the associated principal content of the cover image has been tampered with. Hence our requirement is to make such a type of fragile watermark that is highly fragile and sensitive enough for the smallest change even a single bit.

6.1.1 Fragile Watermark as a Hash Function for Images

A hash function is a mathematical function that converts a numerical or textual input value into another compressed numerical or textual value. The input to the hash function is of arbitrary length but the output is always of fixed length. Values returned by a hash function are called a message digest or simply hash values. There is a huge difference between the length of the input and output values for a hash function, and it is very difficult to generate exactly a one to one (input to output) hash function, but we still can generate a hash function which can map one input with as few as possible output values. If any tampering is done on the input numerical or textual information then its associated hash value will also be changed, hence in this way one can judge the integrity of the input value. A fragile watermark is an alternative method for a hash function in case of images, because it would be very time consuming process to apply a hash function to each pixel intensity of the cover image. Fragile watermarks play the same role as a hash function in order to check the integrity or authentication of the given image. In the case of image processing, the word integrity and authentication are used interchangeably, so do not be confused. Both deal with the tampering of content.

6.1.2 Fragility of a Fragile Watermark

In the previous subsection, we saw that a fragile watermark plays the role of hash functions for images, so it must follow all the properties of hash function.

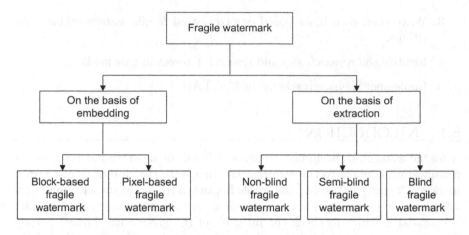

FIGURE 6.1   Classification of fragile watermark.

The generated watermark must be strongly mapped to the principal content of the cover image so that if any tampering is done on the watermarked image, it must be reflected by its fragile watermark. By this way we can simply mark the tampered and untampered pixels of the given image. Mapping of the fragile watermark to the principal content of the given cover image is directly proportional to the fragility or sensitivity of that watermark for various intentional or unintentional attacks. High fragility tends towards a good fragile watermarking approach. An ideal fragile watermark is consistently sensitive enough for even a single bit alteration of any pixel in the watermarked image.

Points worth remembering:
A good fragile watermark scheme has high fragility even for a single bit alteration.

6.1.3   Types of Fragile Watermark

Categorization of fragile watermark, can easily be understood from Figure 6.1. Here we can see that a fragile watermark can be broadly classified in two ways:

1. On the basis of the embedding mechanism.

2. On the basis of the extraction mechanism.

6.1.3.1 On the basis of the embedding mechanism

When a fragile watermark is generated, it must be embedded into the cover image in such way that it can effectively do its assigned job (tamper detection). On the basis of fragile watermark embedding techniques, it is further classified into two categories:

1. **Block-based Fragile Watermark:** As its name suggests, in this category, the watermark is generated for the blocks of a given cover image. First of all the cover image is divided into the non-overlapping blocks of fixed size. After that, the generated watermark is embedded into each block. The main advantage of this type of fragile watermarking scheme is its low complexity because we do not need to process each and every pixel. We treat the whole block as a single pixel. Block-based fragile watermarking schemes are less accurate because if a single pixel of the given block is altered, then the whole block will be treated as an altered block, in spite of having some pixels of that block unaltered. Hence this type of fragile watermark is unsuitable for determining the exact localization of tampering. This scheme is useful when we need to know only whether the given watermarked image in authentic or not.

   If we want to decrease the complexity of the block-based fragile watermarking approach, then we need to take the large size of non-overlapping blocks for the given cover image. But if we want to increase the accuracy of the given scheme, then we need to take the small size of the non-overlapping blocks for the given cover image. Hence there is a trade-off between the complexity and the accuracy in the block-based fragile watermark. In order to make a good block-based fragile watermark, we need to make a moderate size of non-overlapping blocks to obtain good accuracy for tamper detection with less complexity.

2. **Pixel-based Fragile Watermark:** In this category, a watermark is generated using each pixel of the given cover image. For this reason, a pixel-based watermark is most suitable for exact localization of tampering. If the size of the cover image is very large then the complexity of this algorithm is very high. This category of watermark is more applicable in real life applications like medical image authentication, military applications, security of court evidence etc. In the era of advanced processors, some times we can compromise on the time complexity but we cannot compromise on security issues. Hence a pixel-based fragile watermark is preferred for real life applications where security is a major concern.

Points worth remembering:
Block-based fragile watermarks are less complex without exact tamper localization.

Points worth remembering:
There is a trade-off between the block size and accuracy in block-based
fragile watermark.

6.1.3.2   On the basis of the extraction mechanism

At the receiver end, once the watermarked image is received, we need to extract the watermark in order to check the integrity of the watermarked image for any further application. On the basis of watermark extraction, a fragile watermark is further classified into three categories:

1. **Non-Blind Fragile Watermark:** In this category of fragile watermark, we require the original cover image for tamper localization at the receiver end. This class of watermark is highly undesirable because it increases the overhead for storing and transmitting the cover image along with the watermarked image.

2. **Semi-Blind Fragile Watermark:** In this category of fragile watermark, instead of the whole cover image, we require only some side and abstract information of the cover image at the receiver end to detect the tampered region. Sometimes we embed the watermark in some region-of-interest (ROI) of the cover image (see in detail in the next section). In such cases we need to keep track of the embedding locations of the watermark in the cover image. This track record can be treated as side or abstract information of the cover image and it is required at the receiver end in order to detect the tampered pixels. We need to transmit this side information over a secure channel or this information must be encrypted to prevent it from any middle attack.

3. **Blind Fragile Watermark:** This category of fragile watermark is most desirable because it does not require any information related to the cover image at the receiver end during the tamper detection process. The watermark embedding and extraction algorithms are well known to all, so to prevent the unauthorized extraction or embedding of invalid watermarks into the cover image, we need some symmetric key known by only the sender and receiver. In the case of blind fragile watermark, we only need a symmetric key at the receiver end which has been used during embedding of the watermark.

Points worth remembering:
There should be a secure mechanism to transmit the auxiliary information
required for extraction of the watermark at the receiver end.

6.2 GENERATION OF A FRAGILE WATERMARK

In the previous section we went through the basics and background related
to fragile watermarks. Our objective here is to generate a fragile watermark
for a given cover image in order to ensure its integrity. In this section we
will see various concepts which can be applied to generate an efficient fragile
watermark. All those techniques which may be used to generate a fragile
watermark can be referred to as fragile watermarking approaches.

6.2.1 Image-Based Fragile Watermark

Most of the fragile watermarks are nothing but other images which are embedded into the cover image. Since we are dealing with gray scale as well as
color cover images, binary images for fragile watermarks are preferred. We
know that in the binary images, a single pixel can be represented by a single
bit either 1 or 0. Here 0 represents the black intensity whereas 1 represents the
white intensity. Suppose we have a gray scale cover image I of size $M \times N$.
In this type of image, each pixel is represented in eight bit binary form. If
we want to embed an image W as a fragile watermark, then the following
conditions must be satisfied:

1. The dimension of W and I must be the same ($M \times N$).

2. I must be gray scale or a color image whereas W must be a binary
   image.

Since both images have identical dimensions, a bit of each $(x, y)^{th}$ location
of W will be treated as the fragile watermark for the corresponding $(x, y)^{th}$
location of I. Now the question is, where and how should this watermark
information of W be embedded into I? Since each pixel of I can be represented
as eight bit binary form whereas each pixel of W can be represented as a single
bit, a single bit of the $(x, y)^{th}$ location of W can easily be embedded into any
bit of the corresponding pixel of I. Bit selection in a cover image's pixel for
embedding a watermark bit is crucial because the wrong selection of a location
for bit embedding can degrade the quality of the watermarked image as well
as reduce the tamper localization rate. A visual example of an image-based
fragile watermark is shown in Figure 6.2. In this figure, we can see that the
cover image "Lena" is gray scale whereas the watermark image "Baboon" is
binary. Here the binary watermark image is embedded into the first LSB of the

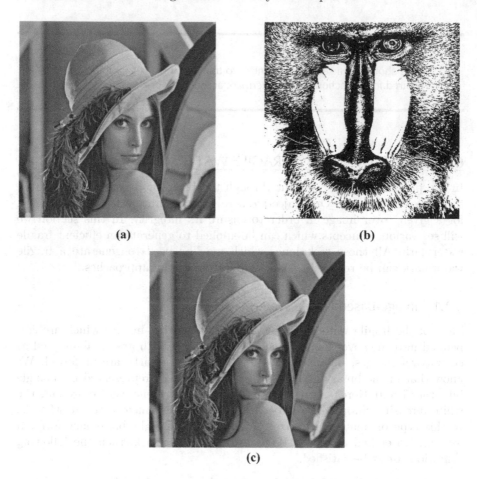

(a) **(b)**

(c)

FIGURE 6.2 Example of an image as a fragile watermark: (a) Original cover image, (b) Binary watermark image, (c) Watermarked image.

cover image. That is why both cover and watermarked images look visually similar and the imperceptibility between both of them is very high in terms of Peak Signal to Noise Ratio (PSNR). In the next section we will see in detail from where and how many bits should be taken for the watermark embedding process.

6.2.1.1 Relation between cover image and image-based watermark

An Image-based fragile watermark is not preferred because most of the time there is no relationship between the cover and watermark images. Let us consider that both images are independent so there is no dependent relationship between the corresponding pixel values of the images. Similarly, the watermark bit is embedded into the kth bit of each pixel of I. In such a scenario, alteration of the watermarked image can be detected if and only if the kth bit of the watermarked image is altered because of any intentional or unintentional attack, because only that bit is treated as fragile watermark bit among all eight bits. If all seven bits of a pixel are altered instead of the kth bit, in that case too, that pixel will be treated as unaltered, because in spite of the change in the seven bits, the watermark bit is safe. Hence an image as a fragile watermark is highly undesirable and misleading. Figure 6.3 shows the inefficiency of image-based fragile watermarks. In this image, a binary watermark shown in Figure 6.3(b) is embedded as a fragile watermark and the watermarked image is shown in Figure 6.3(a). A region-of-interest-based attack is carried out on the watermarked image which is shown in Figure 6.3(c). Since in this example, the watermark is embedded into only the first LSBs of all pixels, this embedded watermark can only detect the tampering of the first LSB of the watermarked image. Figure 6.3(d) shows the tamper detected regions. Since there is no relationship between the cover and watermark images, we are unable to do exact localization of the tampered region.

Points worth remembering:
A binary image-based fragile watermark is highly undesirable because there is no relation between the corresponding pixels of the cover and watermark image.

FIGURE 6.3 Example of tamper localization in the case of image-based fragile watermark: (a) Watermarked image, (b) Binary watermark image, (c) Tampered image, (d) Tamper localization. *(Courtesy of Digital Image Processing, 3rd ed., by Gonzalez and Woods, Pearson/Prentice Hall, 2008).*

MATLAB Code 6.1

```
%MATLAB Code 6.1
% This MATLAB code is for binary watermark embedding in gray cover image.
clc
clear all
close all
file_name='lena.bmp';
cover_image=imread(file_name);% Reading the cover image.
file_name='baboon.bmp';
watermark=im2bw(imread(file_name));% Reading and converting the watermark image in binary.
[Mc Nc]= size(cover_image);% Mc= No. of rows, Nc= No. of Columns.
watermark=imresize(watermark,[Mc Nc]);% Make the size of cover and watermark image equal.
% set the 1st LSB of cover_image(ii,jj) to the value of the MSB of watermark(ii,jj)
watermarked_image=cover_image;
for ii = 1:Mc
    for jj = 1:Nc
        watermarked_image(ii,jj)=bitset(watermarked_image(ii,jj),1,watermark(ii,jj));
    end
end
% write to file the two images
imwrite(watermarked_image,'lsb_watermarked.bmp','bmp');
% display watermarked image
figure(1)
imshow(watermarked_image,[])
title('Watermarked Image')
```

MATLAB Code 6.2

```
%MATLAB Code 6.2
% This MATLAB code is for watermark extraction from the watermarked image
clc
clear all
close all
file_name='lsb_watermarked.bmp';
watermarked_image=imread(file_name);% read in watermarked image
[Mw Nw]=size(watermarked_image); % determine size of watermarked image
% use lsb of watermarked image to recover watermark
for ii = 1:Mw
    for jj = 1:Nw
        watermark(ii,jj)=bitget(watermarked_image(ii,jj),1);
    end
end
figure(1)
imshow(watermark,[])% scale and display recovered watermark
title('Recovered Watermark_32jh')
```

6.2.2 Self-Embedding Techniques

In the previous subsection we saw that an image-based fragile watermark is not a good option because of its lack of relationship with the cover image. Hence we need a technique which can generate a fragile watermark that strongly relates to the cover image. Self-embedding is a technique which generates a very efficient fragile watermark. These types of watermarks have no visual meaning, which means this watermark is meaningless in nature unlike the previous one.

Through self-embedding techniques, one can make one's own fragile watermarking approach. Self-embedding techniques can be further categorized in two ways:

1. **Pixel-based self-embedding**

2. **Block-based self-embedding**

Pixel-Based Self-Embedding:

In this case, our objective is to make an algorithm which generates a unique sequence of k bits for l bits of each pixel of image I where k and l hold the following conditions:

1. $k \ll l$.

2. There should be some algorithms, when applied on l bits of a pixel of I that must provide a unique sequence of k bits.

3. Different patterns of l bits must generate different but unique sequences of k bits.

4. This is the concept of a hash function for images.

It is a very difficult task to generate k unique bits for each set of l bits. It is just like a hash function. We cannot ideally generate one-to-one mapping of k and l bits because k is smaller than the l. But our objective is to create an algorithm which most of the time gives unique k patterns for different patterns of l bits. The dependency between l and k must be very high such that a variation in a single bit of l would change the value of k.

Here k means the fragile watermark bits and l means the principal bits of the corresponding pixel. If we are using a gray scale image as cover then its pixel can be represented in eight bit binary form. k and l are related to each other by the relation $l = 8 - k$. Figure 6.4 shows the relation between l and k. We can choose any combination of l and k shown in Figure 6.4 as per the requirement of tamper detection accuracy and imperceptibility between cover and watermarked image. A larger value of l gives good imperceptibility but a low tamper detection rate whereas a larger value of k gives a good tamper detection rate with low imperceptibility (explanations are given in upcoming sections).

6.2.2.1 Relation between cover image and self-embedding watermark

Unlike an image-based watermark, here the generated watermark is strongly related to the cover image. In the self-embedding technique, the watermark is generated by using l bits of each pixel of the cover image. Our objective is to develop a relation which may be dependent upon the locations, values, spatial positions etc of all l bits of any pixel. These self-generated relations will provide k bits as output, and these bits are nothing but the fragile watermark for the corresponding pixel. Since k bits are the resultant output of any

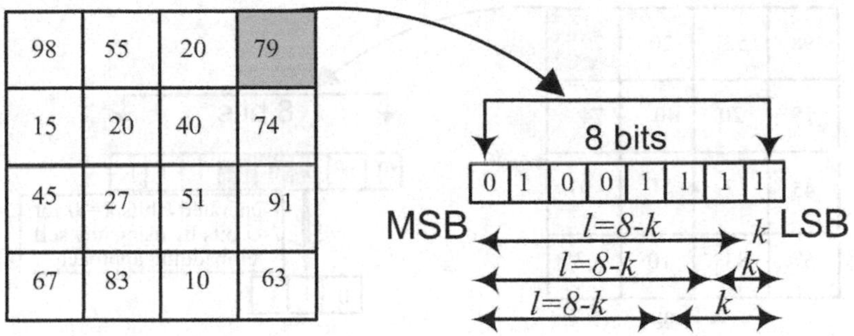

FIGURE 6.4 Understanding of relation between k and l.

bitwise logical operations on l bits as operand, hence if there is any change in l bits, the output k bits most likely are to be changed. This feature is called fragility of k. We need to create a type of self-embedding technique in which small changes on l must be reflected on k. That is why we can say that in self-embedding, there is a strong coupling between the cover image and watermark bits, hence this technique is more preferable.

Points worth remembering:
Self-embedding is the most suitable fragile watermark generation approach because it is completely based on user-defined algorithms and it is more accurate because of its tight coupling between the cover and watermark bits.

6.2.3 Example of Self-Embedding Techniques

Let I be an image of size $M \times N$ and let a pixel of $(x, y)^{th}$ location in I be denoted by $P(x, y)$. If we are dealing with gray scale cover images then a pixel can be represented as eight bit binary form and $P_b(x, y)$ represents the bth bit of the pixel where $b = 1$ to 8 (1 represents LSB and 8 represents MSB). The decision making point here is the selection of an appropriate value for k. Once k is finalized then all pixels of I will generate k watermark bits. In the next section we will see in detail why and how many bits should be allocated for a watermark per pixel i.e. k. The number of k is directly related to the tamper detection rate and quality of the watermarked image. A larger value of k increases the tamper detection rate but decreases the imperceptibility of the watermarked image (See details in next section). Assume that we have

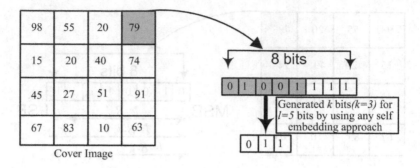

FIGURE 6.5 Example of pixel-based self-embedding approach.

decided to generate k bits as the watermark out of eight bits per pixel. In this case, we will do the following operations for a self-embedding technique:

Step 1. Convert each gray value of each pixel into eight bit binary form. If the pixel intensity $P(x, y)$ of $(x, y)^{th}$ location is converted into eight bit binary form, then $P_b(x, y)$ represents the bth bit of the pixel where $b = 1$ to 8 (1 represents LSB and 8 represents MSB).

Step 2. Convert k LSBs of each pixel into 0, where k is the user-defined watermark bits per pixel.

Step 3. Apply any self-generated self-embedding approach on $l = 8 - k$ bits of a pixel to output in k bits. These operations will be based on either the values of l bits or their locations or any other credentials of l bits.

We can create a single self-embedding approach to generate k watermark bits as output or we can generate k different self-embedding approaches, to generate one watermark bit, hence k approaches generate k bits together. The second approach is more preferable because in this case we can tightly couple the l bits with their credentials and the probability of changing any k watermark bits with respect to change in l bits of that pixel will be higher in this case. Because all k watermark bits are generated by different self-embedding methods, we have k chances to detect the alteration. Th fragility of k bits will be very high in this case. Figure 6.5 shows the conceptual view of a pixel-based self-embedding approach. In this figure we can see that $l = 5$ MSBs are chosen for watermark bit generation per pixel. These bits will be given to any pixel-based self-embedding approach as input which gives three watermark bits($k = 3$) per pixel as output.

Here we provide some examples of self-embedding approaches. The reader can generate his own improved approach on the basis of these concepts.

1. **Self-embedding approach based on l bit values:**

$$W_b = \sum_{i=k}^{7}(P_i(x,y) \oplus P_{i+1}(x,y))mod2 \qquad (6.1)$$

Explanation: In this method we can see that each ith bit (starting from k to 7) is bitwise XORed with its consecutive $(i+1)^{th}$ bit and summed with its previous value. Once the XOR operation is completed, we do a modulo 2 operation in order to make it either 0 ot 1. This final generated bit W_b is the watermarked bit.

This is the a very trivial example of self-embedding technique. Here we are just associating the bits of the cover image. In the aforementioned example, we can see that if any bit value from $l = k$ to 8 gets changed, then most of the time output bit value W_b will also be changed. We can understand it with an example:

Example 6.1: Let $P(x,y) = 57$ and according to step 1 its eight bit binary representation is $P(x,y) = 00111001$. Now as per step 2, if $k = 1$ then convert the first LSB to 0 and the resultant $P(x,y)$ will be 00111000. If we apply the aforementioned self-embedding approach in order to generate a single fragile watermark bit for this pixel, then equation 1.1 will provide $(0+1+0+0+1+0)mod2$ i.e 0. It means $W_b = 0$. Suppose during transmission intentionally or unintentionally the first MSB of the pixel is changed and so $P(x,y) = 10111000$. If we again apply the aforementioned self-embedding approach, then the result will be $(1+1+0+0+1+0)mod2$ i.e 1. It means that we have related the k and l bits of a pixel such that a slight modification on the l bits can be tracked. But this is not the ideal approach because most of the time it will fail. For example suppose during transmission intentionally or unintentionally the second MSB of the pixel is changed and so $P(x,y) = 01111000$. Then the result will be $(1+0+0+0+1+0)mod2$ i.e 0 by the aforementioned self-embedding approach. It shows that sometimes we are able to track the changes and sometimes we are not. Hence we have to create a type of algorithm which has a high probability of tracking the changes.

2. **Self-embedding approach based on l bit's positions:**
   Each pixel has its own unique spatial location on an image. We can simply use this concept to connect the pixel with its row, column or diagonal locations. By doing logical operations, we can convert the desired number of fragile watermark bits W_b. If the pixel value will be changed, then its relation with its position will also be disturbed and hence the generated W_b will be changed. By this way we can easily track the altered pixel.

3. **Self-embedding approach based on other features of pixels**

in the cover image:
The above mentioned approaches are just examples. One can make one's own self-embedding approach using some of the other relevant features of the pixel or image. For example we can also take the hamming weight of a pixel to generate the W_b or we can associate the pixel with its N_4 or N_D neighbors. One can map the pixels with other random numbers known by only the sender and receiver. There may be a number of ways for self-embedding approaches. This is an area of research to generate those types of self-embedding methods which are highly fragile with respect to very few changes in the images.

4. **Here one more question arises:** Where should the generated W_b be embedded in the cover image? In the second step we deliberately converted the K LSBs to zero. These W_b bits will be embedded in those zero LSBs only. We will discuss watermark embedding in more detail in next section.

Points worth remembering:
Watermark bits W_b can be generated by applying any relation or association on the cover image bits (based on their value, position, neighbour or any other feature).

Block-Based Self-Embedding: In this type of self-embedding, our attention is concentrated on blocks rather than the pixels. This type of procedure is used when we are more sensitive to the quality of the watermarked image and hence we have much less space for watermark embedding in the cover image. The following procedure is applied for this category of self-embedding:

1. First of all the cover image I of size $M \times N$ is divided into non-overlapping blocks of size $m \times n$. Here $m << M$ and $n << N$.

2. Now each block will be treated as small image and pixel-based self-embedding will be applied on each block.

3. After applying on each block, do the following operation:

$$W_b = \sum_{1}^{m \times n} (Pixel\ based\ self-embedding\ expression) mod2 \quad (6.2)$$

Explanation: The output bits of the pixel-based self-embedding technique applied on each pixel of a block will be added. So after adding the total $m \times n$ bits, we apply a modulo 2 operation on that value in order to get a single W_b as output.

4. Here the same question will arise: Where will the generated W_b be embedded in the block? These W_b bits of every block will be embedded in the selected pixel's LSB. We will see watermark embedding in more detail in the next section.

Points worth remembering:
Block-based self-embedding techniques give better quality of watermarked image with lower tamper detection rate, because W_b is highly compact.

Points worth remembering:
A more general and less complex fragile watermark has yet to be developed.

Figure 6.6 shows the example for a block-based self-embedding approach. In this example, the given image is divided into non-overlapping blocks of size 2×2 and the $k = 3$ watermark bits for a single block are generated by using $l = 5 \times 4$ i.e. 20 bits. Here we are taking five MSBs from each of the four pixels in each block.

6.2.4 Significance of the XOR Operation in Self-Embedding

Most of the self-embedding approaches use XOR based bitwise operations. The logic behind it is its reversibility property and the tendency to give maximum variations in its truth table (two times 1 and two times 0 for XOR operations of four different combinations of 1 and 0), which is not there in other logical operations. If P and Q are two bits then the XOR operation satisfies the following conditions:

1. $P \oplus Q = Q \oplus P$.

2. $P \oplus P = 0$.

3. $P \oplus R$ is random where R is a random bit.

4. $P \oplus Q = S$ implies $S \oplus Q = P$.

Here the third and forth properties are essential for fragile watermarking. According to third property, we can encrypt the watermark information by

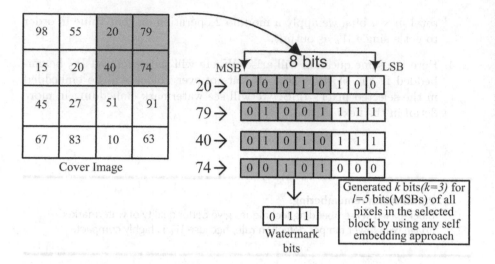

FIGURE 6.6 Example of block-based self-embedding approach.

doing its XOR with any random matrix. Similarly the fourth property is called reversibility, which can be used to encrypt and decrypt the watermark information using a symmetric key (see in the next topic).

6.2.5 Fragile Watermark with Symmetric Key

It is a well known fact that once algorithms come into the public domain, they are no longer private. If an attacker knows the algorithm, he can intentionally alter the watermark from the embedded bits in such a way that in spite of carrying out his intentional malicious attack on the watermarked image, it will be treated as unaltered. Hence to prevent this type of mis-happening, a symmetric key can be used at the sender end to encrypt the generated watermark information before embedding. Since we have a huge number of pixels to encrypt, that is why for simplicity we can generate a random matrix R of the same size as the cover image with the help of the symmetric key. The symmetric key will be treated as a seed value for that random matrix. All generated fragile watermark bits for all pixels are arranged in a matrix form and that matrix is denoted by W_m. Both matrices will be the same size as the cover image. After getting the binary random matrix we can simply apply the XOR operation on the corresponding bits of both matrices.

$$W_b = R \oplus W_m \tag{6.3}$$

Here W_b is the encrypted matrix for the watermark bits. Since the size of the W_b will be the same as the cover image, it can be directly embedded into the corresponding pixels of the cover image. Since the key used is symmetric, it will also be used at the receiver end for decryption of the watermark matrix. Because of the reversibility property of the XOR operation, the original watermark matrix can be achieved by just doing the bitwise XOR operation between the matrix W_b and regenerated random matrix R at the receiver end generated with the same symmetric key .

$$W_m = R \oplus W_b \tag{6.4}$$

Points worth remembering:
Encryption of a generated fragile watermark with the help of a symmetric key enhances the security of the approach.

Points worth remembering:
XOR is the most suitable logical operation for self-embedding because of its property of reversibility and tendency to give the maximum output.

MATLAB Code 6.3

```
%MATLAB Code 6.3
%MATLAB code for random matrix generation using secret key.
sk=input('Enter the Secret Key for Authentication');
rand('state',s1);% Fixing the seed value for matrix.
R=rand(Mw,Nw); % Generate random matrix of size Mw x Nw, same as watermark (values
    between 0 to 1)
R1=round(R);% change the values either 0 or 1.
imshow(R1,[])% Display the random matrix as image.
```

MATLAB Code 6.4

```
%MATLAB Code 6.4
%MATLAB code for generation of encrypted watermark using secret key.
W=imread('watermark.bmp'); % Read the binary watermark.
[Mw Nw]= size(W); % Calculating the size of W
W= logical(W); % Convert the pixels in logical 1 or 0 form;
sk=input('Enter the Secret Key for Authentication');
rand('state',s1);% Fixing the seed value for matrix.
R=rand(Mw,Nw); % Generate random matrix of size Mw x Nw, same as watermark (values
    between 0 to 1)
R1=round(R);% change the values either 0 or 1.
R2= logical(R1); % Converting pixel as logical 1 or 0;
Enrpt_watermark= xor(R1,W); % Do XOR logical operation between random matrix and
    watermark.
imshow(R2,[])% Display the encrypted watermark.
```

6.2.6 Watermark Generation for Color Images

A color image can be treated as the combination of three color planes. For example if we are dealing with the RGB color model, then a color image is the combination of three color planes red, green and blue. An individual color plane can itself be treated as a gray scale image because the value of its intensities ranges from 0 to 255. Now we can generate and embed the watermark bits for each plane individually by the aforementioned methods and combine them in order to create a color watermarked image. Some times to enhance the security, we secretly choose a single or double color plane out of three to embed the watermark. Te identification clue for chosen color planes are securely transmitted to the receiver so that he can extract watermarks from that embedded plane.

Points worth remembering:
A watermarked color image is nothing but the combination of three water-marked gray scale images.

6.3 EMBEDDING OF A FRAGILE WATERMARK

In order to get high fragility, we need to develop a good embedding mechanism for the fragile watermark. If the embedding process is easy, then the extraction process will also be easy. In upcoming sections we will deal with the process related to embedding and extraction of fragile watermarks so all schemes which deal with these processes can be treated as fragile watermarking schemes. There are various factors such as imperceptibility between watermarked and cover images, fragility etc that must be kept in mind while making a good fragile watermarking scheme.

6.3.1 Domain Selection

An image can be represented in two domains viz the spatial domain and frequency domains. Each domain representation has its own benefits. We need to choose the domain which will be more suitable for fragile watermark embedding. First of all we should learn about both domains.

6.3.1.1 Spatial domain

In this domain we directly deal with the intensities of the pixel. We can directly apply additive, multiplicative or subtractive operations on intensities of the pixels and reflections of those operations can be observed in the corresponding images. Pixel-based operations in this domain are very easy to understand as well as to implement.

6.3.1.2 Frequency domain

In this domain we do not apply direct operations on the intensities of pixels. Here we take the transformations (Fourier, DCT, DWT etc) of the images. After getting the transformation coefficients, we apply the operations on them.

6.3.1.3 Which domain is suitable for a fragile watermark?

We need to choose a domain which is perfect for fragile watermark embedding for valid reasons. As we know that a good fragile watermark can be judged by its level of fragility or sensitivity against various minor and major attacks. Hence we have to choose the domain for embedding which is sensitive enough against the smallest distortion on images. In the case of the frequency domain, transformation coefficients for a particular location of the pixel are calculated by all pixels of the images. That is why most of the transformations are reversible in spite of missing some principal content of the images due to major attacks. Hence we can say that the frequency domain is a kind of robust domain that can tolerate minor and major attacks and hence is unsuitable for fragile watermarking. But still this domain is more suitable for other kinds of watermarking schemes, i.e., robust watermarking. The pixels are more vulnerable than the transformation coefficient of the pixel to any minor distortion. Even a small addition of noise can change the original values of the pixels. Hence we can conclude that the spatial domain is more fragile and sensitive against minor distortion on images (as it directly deals with pixels) and is well suited for fragile watermark embedding.

Points worth remembering:
A good fragile watermark in the frequency domain still needs to be generated.

6.3.2 Bit Plane Slicing

In the previous section, we saw why the spatial domain is beneficial in order to embed the fragile watermark. Now the question arises of which bit plane is most suitable for watermark embedding. In MATLAB the default format of an image is 'uint8' i.e. unsigned eight bit integer format. It means a pixel intensity can be represented by eight bits and total number of intensities will be 256 (2^8) ranging from 0 to 255. The visual representation of 0 intensity is black and 255 is white. Bit plane slicing is a process in which we separate

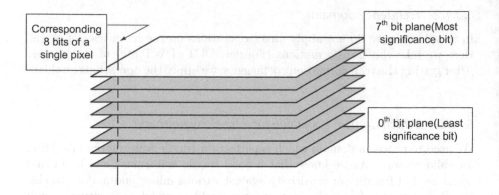

FIGURE 6.7 Concept of bit plane slicing of an 8 bit image.

individual bits of all pixels of an image and create eight binary images as shown in Figure 6.7. We can understand it through simple example 6.2.

Example 6.2 Let $I = \begin{bmatrix} 128 & 36 \\ 29 & 10 \end{bmatrix}$ be an image of size 2×2. The eight bit binary representation of the image is $I_b = \begin{bmatrix} 10000000 & 00100100 \\ 00011101 & 00001010 \end{bmatrix}$. Now eight binary images can be formed by extracting the corresponding bits of all pixels. If the image formed by i^{th} bit is represented by I_i where $i = 1\ 8$ then:

$$I_1 = \begin{bmatrix} 0 & 0 \\ 1 & 0 \end{bmatrix}, I_2 = \begin{bmatrix} 0 & 0 \\ 0 & 1 \end{bmatrix}$$

$$I_3 = \begin{bmatrix} 0 & 1 \\ 1 & 0 \end{bmatrix}, I_4 = \begin{bmatrix} 0 & 0 \\ 1 & 1 \end{bmatrix}$$

$$I_5 = \begin{bmatrix} 0 & 0 \\ 1 & 0 \end{bmatrix}, I_6 = \begin{bmatrix} 0 & 1 \\ 0 & 0 \end{bmatrix}$$

$$I_7 = \begin{bmatrix} 0 & 0 \\ 0 & 0 \end{bmatrix}, I_8 = \begin{bmatrix} 1 & 0 \\ 0 & 0 \end{bmatrix}$$

One can understand the significance of bit plane slicing from Figure 6.8 where the visual contribution of each plane is given. Figure 6.8 (a) is the original gray scale image which can be represented in eight bit planes. Figure 6.8 (b) to (i) are the first to eight bit planes, respectively. Here we can see that we cannot infer any information related to the visual content of the cover image from the LSB plane. As we move towards MSB planes then visualization of cover image information becomes more clear. It means LSB planes contain the detailed or texture information of the cover image whereas MSB planes contain structural or edge information of cover image. By this way it is clear

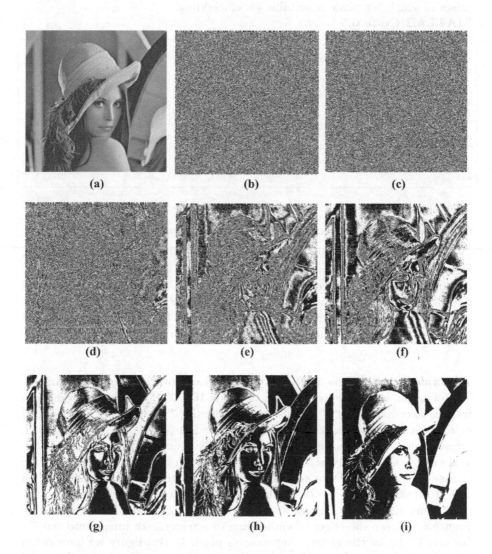

FIGURE 6.8 Example of bit plane slicing of an 8 bit gray scale image:
(a) Original gray scale image, (b), (c), (d), (e), (f), (g), (h) and (i) are
the binary images generated by the 1st, 2nd, 3rd, 4th, 5th, 6th, 7th and
8th bit planes respectively.

that alteration in LSB planes cannot be detected easily due to its randomness whereas alteration in MSB planes can be easily identified. Now the question arises of which bit plane is suitable for embedding.

MATLAB Code 6.5

```
%MATLAB Code 6.5
% This MATLAB code is for demonstration of bit plane slicing
clc
clear all
close all
%Bit slicing
i=imread('lena.bmp'); % Read the image
for b=1:8
    a=bitget(i,b);
    a=logical(a);
    figure(b);
    imshow(a);
end
```

6.3.2.1 Which bit should be chosen for embedding?

From figure 6.8 we saw that LSBs (from right to left) store detailed information whereas MSBs store information of edges. If we embed the watermark bit into LSBs it reflects only a very minor change in the watermarked pixel which cannot be visualized by anybody whereas if we embed into MSBs, a very big change in intensity is reflected which can be easily observed by the human visual system. This can be understood from example 6.3.

Example 6.3 Let $P = 128$ be a pixel intensity of the given secret image. The eight bit binary representation of 128 is 10000000. Suppose we embed a fragile watermark bit "1" in this intensity at the 7th bit (from right to left) of P, then the watermarked intensity will be 192 (11000000) which is much larger than the original one. It can also easily be tracked by our human visual system. Cryptanalysis is applied quite easily in this type of scenario. If we choose 1st bit (1st LSB) for embedding then the watermarked intensity will be 129 (10000001). There is not much difference between the visual appearance of 128 and 129, so the human visual system cannot identify the embedding. Figure 6.9 shows the effect of embedding of a watermark image into the 7th bit and 1st bit of the given cover image's pixel. In this figure we have taken a gray scale image of size $M \times N$ as the cover image and a binary image of the same size as the watermark image. Figure 6.9(d) shows the embedding of watermarked information into the 7th bit. In this case distortion can easily be tracked whereas when we embed the watermark in the 1st bit shown in Figure 6.9(c), it cannot be tracked.

Points worth remembering:
Least Significant Bits (LSBs) are the most suitable place for fragile watermark
embedding.

6.3.2.2   How many bits should be chosen for embedding?

Till now we have argued that embedding in LSBs is more secure and less vulnerable to cryptanalysis. The aforementioned examples and figures support
this claim. Now the problem is how many bits per pixel should be dedicated
for fragile watermark embedding. Let us say we are dedicating the first LSB
of each pixel for watermark embedding. Assume that this watermark bit is
unrelated (see in the previous section about related and unrelated fragile watermarks) to the rest of the seven bits of the pixel. Hence tamper detection will
only be possible when this first bit is changed. If the rest of the seven bits are
changed except the first one, that pixel will still be treated as unaltered, because determination of tampering is done only with the help of the embedded
watermark bit and in this case that bit is still safe. A single bit can be altered
with 50% probability. Bit 0 can be 1 or bit 1 can be 0 during tampering. It
means that if we embed only a single bit as a watermark in each pixel then
we can only achieve at max 50% tampering detection. To increase the tamper
detection rate we need to embed more bits as fragile watermark in each pixel.
As the number of bits of fragile watermarks in each pixel increases the tamper
detection rate also increases. But this is not a good practice for developing a
good fragile watermarking scheme. We will see in the next section why this is
not a good practice. Figure 6.10 shows a progressive degradation of a watermarked image with respect to the quantity of the watermark bit embedding.
Here we can see that as we increase the number of embedded bits per pixel
in the watermarked image, we can experience progressive degradation in the
watermarked image. The conclusion can be drawn that the degradation of the
image is directly proportional to the number of altered bits per pixel.

Points worth remembering:
As the number of bits of fragile watermark per pixel increases, the tamper
detection rate also increases.

FIGURE 6.9 Effect of the watermark on cover image on basis of sequence number of the bit chosen for embedding: (a) Original gray scale image, (b) Binary watermark image, (c) Watermarked image with 1st bit plane embedding, (d)Watermarked image with 7th bit plane embedding.

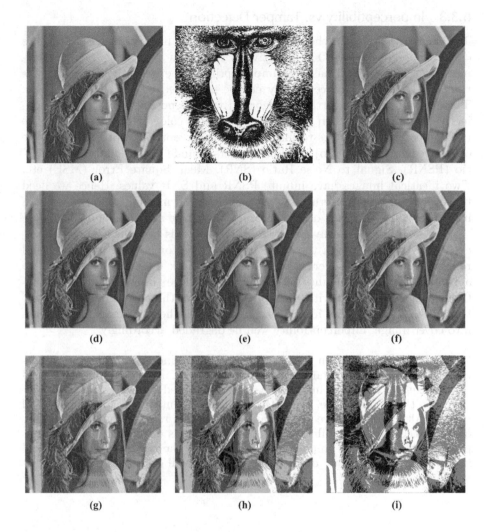

FIGURE 6.10 Cumulative progressive effect of the watermarks on a cover image on the basis of successive bits chosen for embedding: (a) Original gray scale image, (b) Binary watermark image, (c) Watermarked image with 1st bit plane embedding, (d) Watermarked image with 1st and 2nd bit plane embedding, (e) Watermarked image with 1st, 2nd and 3rd bit plane embedding, (f) Watermarked image with 1st, 2nd, 3rd and 4th bit plane embedding, (g) Watermarked image with 1st, 2nd, 3rd, 4th and 5th bit plane embedding, (h) Watermarked image with 1st, 2nd, 3rd, 4th, 5th and 6th plane embedding, (i) Watermarked image with 1st, 2nd, 3rd, 4th, 5th, 6th and 7th bit plane embedding.

6.3.3 Imperceptibility vs. Tamper Detection

The similarity between the original cover image and watermarked image is called imperceptibility. Our objective is to design a type of fragile watermarking scheme which has high imperceptibility between the cover and watermarked image. Imperceptibility can be measured by various parameters. These parameters are further categorized with respect to subjective and objective parameters. When our cover images are color or gray scale then we use subjective parameters to measure the imperceptibility, otherwise we use objective parameters. Subjective parameters include Peak Signal to Noise Ratio (PSNR), Signal to Noise Ratio (SNR), Mean Square Error (MSE) etc. Two identical images have infinite PSNR and SNR values. Hence we need to develop a fragile watermarking scheme keeping in mind that the value of imperceptibility in terms of any parameter must be as high as possible.

You might wonder why we suddenly started this topic or what is the relevancy of this topic with the previous one? This topic is highly related and relevant because whenever we embed a watermark into the cover image we actually degrade the quality of the image. Because the watermark information has nothing to do with the visual content of the cover image, it can be treated like noise. As we increase the quantity of the watermark information into cover image, imperceptibility well be degraded accordingly.

In the previous section we saw that in order to increase the tamper detection rates we needed to embed fragile watermarks in greater number of bits of the pixels of the cover image. But when we embed the watermark information into the cover image, the actual visible content of the cover image will be lost. Although we increase the tamper detection rate, but unknowingly we decrease the imperceptibility. It means that there is a trade-off between tamper detection rate and imperceptibility between the cover and watermarked image. When we increase one's value, the other will decrease.

Points worth remembering:
There is a trade-off between the tamper detection rate and imperceptibility (between cover and watermarked image).

Embedding of fragile watermark can be done in two ways:

6.3.3.1 Block-based embedding

In this case, we first divide the cover image into non-overlapping blocks. Let $m \times n$ be the size of a block. It means that this block will contain $m \times n$ number of pixels and total $m \times n \times 8$ number of bits, if eight bit representation for a single pixel is used. Now secretly some fixed bit spots from $m \times n \times 8$ bits are chosen to embed the fragile watermark. Behind the block-based embedding, our objective is to improve the imperceptibility between the cover and watermarked image. In block-based embedding we summarize the watermarked information of $m \times n$ pixels into a few bits, so that when we embed it, fewer changes are required. Since in this technique mapping of $m \times n \times 8$ bits is done with very few watermarked bits, it is less effective in terms of tamper detection but more effective in terms of imperceptibility. This approach will be more suitable if we choose the bit locations for embedding the watermark dynamically. This means each non-overlapping block can have different locations for watermark embedding. But keeping track of these selected locations and sending them to the receiver end for proper watermark extraction and tamper detection is again very challenging and a difficult task. All approaches which use block-based embedding are called block-based watermarking approaches. Figure 6.11 demonstrates an example for block-based embedding. In this example the cover image is divided into 2×2 non-overlapping blocks. For each block, we are generating $k = 4$ bits using any block-based fragile watermarking approach. The value of k can vary as per the algorithm. Now we have to embed these four watermark bits into the corresponding blocks. There are two ways to embed the watermarked bits. We can either embed the bits of the watermark into the corresponding pixels of that block sequentially or all bits can be embedded into any single randomly chosen pixel of that block. In Figure 6.11, we can see that each bit is embedded into the corresponding pixel of the block hence, all four pixels may be modified. On other hand, all four bits can be embedded into a single randomly selected pixel (i.e., 74 in this example). Block-based fragile watermarks can also be embedded into the pixels of other blocks. At the receiver end by using one-to-one mapping, we can extract the watermark.

Points worth remembering:
Selection of appropriate dynamic bit locations in a non-overlapping block of size $m \times n$ for watermark embedding and sending its information at the receiver end secretly for tamper detection is a challenging task.

6.3.3.2 *Pixel-based embedding*

In this case, all pixels of the cover image are used for watermark embedding. Let $M \times N$ be the size of an image. It will contain $M \times N \times 8$ bits, if eight bit representation is used for a pixel. We reserve a fixed number of bits, let's say

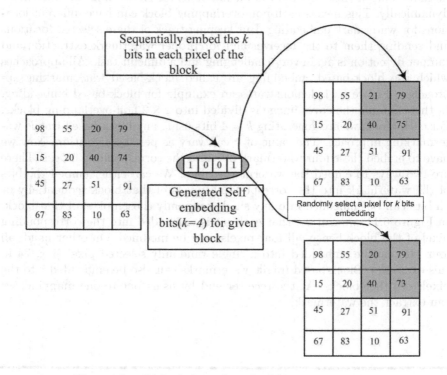

Sequentially embed the k bits in each pixel of the block

Generated Self embedding bits($k=4$) for given block

Randomly select a pixel for k bits embedding

FIGURE 6.11 Example of block-based embedding.

k for each pixel to embed the watermark. Hence the total number of bits for cover information will be $M \times N \times (8 - k)$ and the total number bits used to embed the cover information will be $M \times N \times k$. k bits of a pixel can hold their own principal cover information of $8 - k$ bits of the cover information of any other mapped pixel. From a security point of view, if k bits of a pixel contain the watermark information of $8 - k$ bits of other pixels, then this will be more secure, because this mapping can only be known by the sender and receiver. Hence no one can extract or change the watermark bits in between deliberately. As we know that for better performance, these k bits must be LSBs of the pixels. A larger value of k results in good tamper detection but bad imperceptibility. By this approach, alteration of a single pixel can be traced, which is why it is the most preferable approach. All approaches which use pixel-based embedding are treated as pixel-based watermarking approaches. Figure 6.12 demonstrates the basics of pixel-based embedding. Here $k = 3$ fragile watermark bits are generated for each pixel. In this example, we have shown the embedding for the intensity value 79. Using five MSBs of 79 and any pixel-based self-embedding fragile watermarking approach, we generate three watermark bits. These three bits are embedded into the first three LSBs of the same pixel. Now the updated watermarked pixel of 79 will be 75. The same procedure will be applied for all pixels of the cover image in order to generate the watermarked image.

Points worth remembering:
The selection of an appropriate number for k is a very challenging task because it creates a trade-off between tamper detection and imperceptibility.

Points worth remembering:
Embedding of k bits fragile watermark of a pixel into k bits of an other pixel is preferable. But keeping and transmitting this record at the receiver end is still under research.

6.3.3.3 Region-of-interest (ROI)-based embedding

ROI-based embedding may be a combination of pixel and block-based embedding. Some times, for a cover image, we do not require generating the watermark information for all pixels. Because a cover image can be divided into two parts: principal region and non-principal region. For example an X-ray image containing the information of bones and muscles can be treated as a principal region and all other portions which do not contain any information related to medical diagnosis can be treated as the non-principal region. With

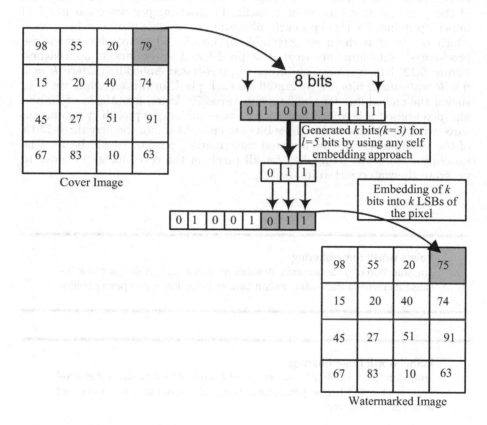

FIGURE 6.12 Example of pixel-based embedding.

the help of a watermark we can save the principal content of the cover image. If we are able to differentiate which portion of the cover image is the principal region and which others are not, then we only generate the watermark bit for the principal region. By this way we can reduce the time complexity of the watermark generation. These generated watermark can be embedded into the non-principal content. By this way we can maintain the imperceptibility of the given image without reducing the quality and without losing a single bit of information in the principal region. Figure 6.13 shows an example for region-of-interest-based embedding. Here a CT head image is used as cover image. In this case, we can easily separate the intensities of the principal and non-principal content of the given CT head image. If the principal region contains P_r number of pixels, then the maximum total number of bits as watermark can be $P_r \times 8$, when we want to save the authenticity of all eight bits of a pixel. Now this information for $P_r \times 8$ bits can be embedded into $(M \times N) - P_r$ number of pixels, which are available in the non principal region. Now the question is how many bits should be taken from each pixel of the non-principal region. We should take only those bits from each pixel of the non principal region for watermark embedding such that the intensity difference between the pixels of the principal and non principal regions must be preserved and could be easily differentiated even after embedding. Let us consider that we are taking k bits from each pixel of non principal content. Then the total number of $((M \times N) - P_r) \times k$ bits are used to store the information of $P_r \times 8$ watermark bits. Here the condition $((M \times N) - P_r) \times k \geq P_r \times 8$ must be satisfied for proper and complete embedding. Otherwise we need to increase the value of k or take less than eight bits of P_r pixels.

Points worth remembering:
Deciding the value of k in order to maintain the relation $((M \times N) - P_r) \times k \geq P_r \times 8$ is a challenging task.

6.4 EXTRACTION OF A FRAGILE WATERMARK

This phase is executed at the receiver end. This phase is very important as it deals with tamper localization. An image may be tampered with either at the time of transmission or at the time of storing that image. Tampering may be further classified into two types:

6.4.1 Unintentional Tampering

In this class of tampering an image may be tampered with during the transmission phase. Most of the time unintentional tampering deals with the insertion of noise due to any small signal fluctuations during transmission. As we know

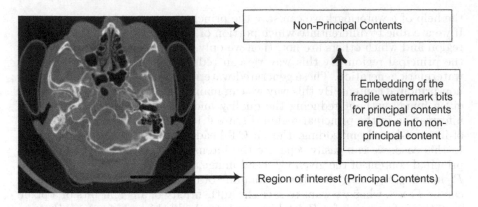

FIGURE 6.13 Example for region-of-interest based embedding of fragile watermark (*Courtsey of Dr. Luca Saba, Radiology Department, University of Cagliari, Italy*).

that any data either image, audio or video is transmitted over the network in the form of bit streams. Hence there is a high probability of toggling the bits of data due to obvious reasons. Noise may belong to the probability density function of any existing noise such as Gaussian, Poison, Salt and Pepper, etc. Figure 6.14 shows an example where the original image is tampered unintentionally by Salt and Pepper noise. As per the property of a good fragile watermarking scheme, even a small change in the bits of the pixel must be detected. Most of the noise directly alters the LSBs of the pixels of the watermarked image. We have learned that LSBs are the most suitable place for fragile watermark insertion. Hence an embedded watermark in LSBs can be destroyed, and this is exactly what we want. Once a watermark is destroyed, we can simply identify the tampered region by the absence of watermarks. Unintentional attacks may also be due to everyday processing task like zooming, cropping, compression etc. Since a fragile watermark is meant for tamper detection, hence any type of tampering must be detected irrespective of its cause.

6.4.2 Intentional Tampering

In this category, an image may be tampered with intentionally at the time of its storage or at the time of transmission by using a man in the middle attack. In this case, the amount of tampering may be very high or very low. In Figure 6.15, an example of an intentional attack is shown where an additional rose is added on the hat of Lena. Tamper detection for this type of attack is very important because these attacks are done in order to mislead individuals. For example, in telemedicine a medical image is transmitted over a network for diagnosis by a remote expert. An attacker may apply a man in the middle

(a)   (b)

FIGURE 6.14 Example of unintentional attack: (a) Original image, (b) Tampered image with salt and pepper noise.

attack and capture the image in order to alter it. Once the image is intentionally altered, it would lead to a wrong diagnosis for the patient. So wrong treatment would be started which could be harmful for the patient. Similarly intentional tampering of images used for court evidence may lead to a wrong conviction. An intentional attack may contain the addition of any object on the image, removal of an object, writing objectionable text on images, etc. A fragile watermark must be capable of identifying all these modifications.

6.4.3 Semi-Fragile Watermarks

On the basis of the tolerance level for the aforementioned attacks, fragile watermarks can be further enhanced to another category called semi-fragile watermark. Semi-fragile watermarks are those watermarks which are treated as robust watermarks for unintentional attacks and are treated as fragile watermarks for intentional attacks. A threshold is used to separate the amount of content-based alteration. If the content-based alteration is less than the threshold value, then it will be treated as an unintentional attack, otherwise as an intentional attack. Semi-fragile watermarks change their role according to the content of the alteration. It is a combination of robust and fragile watermarks. A semi-fragile watermark is made in such a way that it behaves as a robust watermark for unintentional attacks and it behaves as a fragile watermark for intentional attacks.

(a)   (b)

FIGURE 6.15 Example of intentional attack: (a) Original image, (b) Tampered image with addition of a rose on the hat of Lena.

Points worth remembering:
A semi-fragile watermark is the combination of robust and fragile watermarks.

Points worth remembering:
Deciding the threshold value which separates an intentional or unintentional content-based alteration is a very crucial task.

6.4.4 Tamper Localization

Tamper localization is done on the basis of the types of fragile watermarks mentioned in the previous section. A fragile watermark is categorized in three ways(non-blind, semi-blind and blind watermark) on the basis of the watermark extraction mechanism. Tamper localization for each category is a little bit different.

6.4.4.1 Tamper localization for non-blind fragile watermark

As we know that in the case of a non-blind fragile watermark, we also need an untampered version of the watermarked image at the receiver end. Pixel wise comparison is done to localize the tampering. Sometimes there is some tampering which cannot be identified with the human visual system alone, and for that case we require pixel wise comparison of the untampered and tampered watermarked images in this category.

6.4.4.2 Tamper localization for semi-blind fragile watermark

In this category of tamper localization, we do not require an original water-marked image at the receiver end. But here we need some side or abstract information about the watermarked image that contains the locations for watermark embedding. Once we identify the locations where the watermark is embedded, then we simply extract the watermark from the LSBs. Now we again generate the fragile watermark using the same algorithm by which it was generated at the time of embedding. Now we have two sets of watermarks, one is the extracted one and other is the recalculated one. We simply do a pixel wise comparison of these two watermarks and if any mismatch is found between the recalculated one and extracted one then that pixel is marked as the tampered pixel.

6.4.4.3 Tamper localization for blind fragile watermark

In this class, same method as used in semi-blind fragile watermarks, is applied for tamper localization. The only difference is here that we extract and recal-culate the fragile watermark for the entire set of pixels of the image. In this category, we do not need any side information related to locations, because the watermark is embedded in all pixels. We just need a symmetric key at the receiver end from which the watermark is embedded.

The tamper detection result is affected by the way of embedding the fragile watermark, because the watermark can be embedded either by block wise or pixel wise. Hence we can find the accuracy accordingly.

6.4.4.4 Pixel wise tamper detection

If a fragile watermark is embedded through the pixel wise embedding tech-nique, then we can do exact localization of the alteration in the images. This is because each pixel stores its own fragile watermark in its LSBs and if any tampering is done, then the extracted and recalculated LSBs will not match. This type of detection is called pixel wise detection. In Figure 6.16, the im-age of Lena is watermarked with the pixel-based self-embedding technique. An intentional attack is done on the Lena image in which a rose is added on the hat of Lena. At the receiver end, the tamper localization process is per-formed. Figure 6.16 (d) shows the tamper localization result. Here white pixels show the tampered region whereas black pixels show the untampered region. We can see that exact localization of tampered pixels is done in pixel-based self-embedding approach.

6.4.4.5 Block wise tamper detection

If a fragile watermark is embedded using the block wise embedding technique, then we will not be able to determine altered pixels exactly. In this case the image is divided into non-overlapping blocks and the cumulative fragile watermark bits are generated using all pixels of that block. These generated fragile watermark bits are embedded into the parent block or any other mapped block. If any intentional or unintentional tampering is done on any pixel of a block, the entire block will be treated as altered. In Figure 6.17, the image of Lena is watermarked with the block-based self-embedding approach and an alteration is performed on the watermarked image, shown as Figure 6.16. But we can see in Figure 6.17 (d) that the whole block is treated as tampered although it also contains those pixels, which are not actually tampered.

6.4.5 Tamper Detection Parameters

Tamper detection at the receiver end can be judged by two well-known parameters which are the False Acceptance Rate (FAR) and False Rejection Rate (FRR).

False Acceptance Rate (FAR): Sometimes a fragile watermarking scheme detects those altered pixels which have actually not been altered. This scenario is called false acceptance and its rate is called the false acceptance rate. For a good fragile watermarking scheme, FAR must be as low as possible. Figure 6.17 (d) shows some unaltered pixels as altered which is a perfect example of false acceptance.

False Rejection Rate (FRR): If fragile watermarking scheme treats an altered pixel as unaltered, then this scenario is called false rejection and its rate is called the false rejection rate. For a good fragile watermarking scheme, FRR must also be as minimum as possible. Figure 6.16 (d) shows some altered pixels as unaltered (black dots within the white region) which is a perfect example of false rejection.

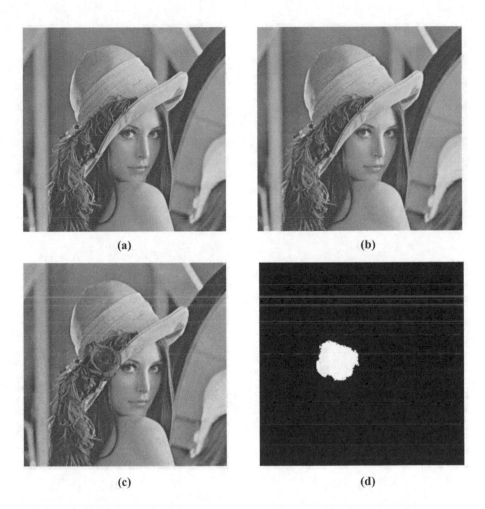

FIGURE 6.16 Example of tamper localization for pixel-based self-embedding approach: (a) Original cover image, (b) watermarked image,(c) Tampered image, (d) Tamper localization.

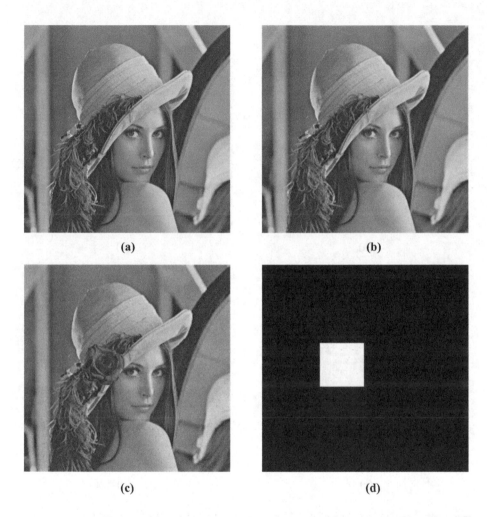

FIGURE 6.17 Example of tamper localization for block-based self-embedding approach: (a) Original cover image, (b) Watermarked image, (c) Tampered image, (d) Tamper localization.

SUMMARY

- *Fragile watermarks* can be used to assist in maintaining and verifying the integrity of their associated cover works.
- Fragile watermarks can be broadly classified in two ways: *on the basis of embedding and on the basis of extraction.*
- On the basis of embedding, fragile watermarking further can be classified in two ways: *pixel-based and block-based fragile watermarks.*
- On the basis of extraction, fragile watermarking further classified in three ways: *non-blind, semi-blind and blind fragile watermarking.*
- *Image-based fragile watermark* is not related to the cover work so it is less efficient.
- The *self-embedding approach* for fragile watermark generation is the best way to generate our own algorithms.
- The self-embedding approach is classified in two ways: *pixel-based self-embedding approach and block-based self-embedding approach.*
- A watermarked *color image* is nothing but the combination of three watermarked gray scale images.
- The *spatial domain* of an image is used for the best processing of the fragile watermarks.
- *LSBs* are the most suitable place for fragile watermark embedding.
- There is a trade-off between the tamper detection rate and imperceptibility between the cover and watermarked image.
- Fragile watermark embedding can be done in two ways: *pixel-based embedding* and *block-based embedding.*
- Attacks can be categorized in two ways: *intentional attacks and unintentional attacks.*
- A *semi-fragile watermark* is the combination of robust and fragile watermarks.
- *Tamper localization* is done on the basis of type of the fragile watermark.
- Tamper localization includes the process of comparison between extracted and recalculated fragile watermarks.
- The efficiency of a fragile watermarking approach is judged by its *False Rejection Rate(FRR) and False Acceptance Rate(FAR).*

CHAPTER 7

Fragile Watermark with Recovery Capabilities in Spatial Domain

CONTENTS

I N the previous chapter, we saw how a fragile watermark can be used to verify the integrity of its associated cover image. Sometimes due to an intentional or unintentional attack, the principal contents of the watermarked image are removed and they must be necessarily recovered for proper understanding of the watermarked image at the receiver end. Till now we have seen fragile watermarks only with detection capabilities which tells us about the location of tampering but we are unable to perform a recovery. In this chapter we will see fragile watermarks with recovery capabilities, which can effectively recover the tampered information of the watermarked image. Fragile watermarks with recovery capabilities will be more desirable when we deal with some sensitive applications like telemedicine, court evidence, military applications, etc.

Hence in this chapter we deal with various theoretical and practical aspects of fragile watermarks with recovery capabilities which deal with the authentication and restoration of the associated cover images. The **Chapter Learning Outcomes(CLO)** of this chapter are given below. After reading this chapter, readers will be able to:

1. Understand different techniques used for recovery in the spatial domain.

2. Compare the pros and cons of different technologies in fragile watermarking with recovery capabilities.

3. Write their own fragile watermarking algorithms with recovery capabilities.

4. Identify the research gap and research interest in this field.

5. Implement their own scheme in MATLAB.

7.1 INTRODUCTION

As we know that a fragile watermark is simply a mark which will be destroyed as soon as any modification is made to the watermarked image. This property of the watermark is suitable for tracing the altered region. Actually this chapter is a continuation of the previous one because once we track the altered region we need to recover it, because there are many applications where we must restore the tampered region. Basically recovery of a tampered region of the watermarked image with the help of fragile watermarks can be done either in the spatial domain or in the frequency domain. Till now we studied the basic differences of both domains. Each domain has its own pros and cons to equip a fragile watermark with recovery capabilities. In this chapter we deal with the fragile watermarks with the recovery capabilities in the spatial domain. As we know that in the spatial domain, we deal directly with pixels hence it is less complex but more vulnerable to attack. In the upcoming sections, we will see the benefits and drawbacks of working with the spatial domain.

7.1.1 Fragile Watermark with Recovery Capabilities

The utility of a fragile watermark may be enhanced if we provide it with recovery capabilities. Till now we have seen that the objective of the fragile watermark is authentication/integrity verification. Once we realize that the integrity of a watermarked image is compromised, then it is the general human tendency to know the actual and original principal content of the watermarked image. So if we embed the recovery information into the cover image along with the authentication information which was only used for tamper localization, then it will be much easier to track as well as restore the tampered content. There are various ways to embed and extract the recovery information in order to restore the image. According to the methods of embedding and extraction, fragile watermarks with recovery capabilities can be further categorized.

7.1.2 Summary Bit Stream: Recovery Information

For authentication/integrity verification of the watermarked image, till now we handled the authentication bit stream. In the previous chapter, we saw various techniques to generate, embed and extract the authentication bit stream. Similarly if we want to recover the tampered information, then we must have some recovery information related to that tampered region. This recovery information will be treated as a summary bit stream. We call it a summary bit stream because if we have n number of bits in principal content and s number of bits for recovery, then $s \ll n$. We need to create an algorithm which generates s bits for n bits of cover image in such a way that either we can perfectly recover the n bits by s bits or we can just get an approximation of n bits at the receiver end. As much as we would be closer to recover n bits by s bits, the algorithm will be more accurate in terms of recovery.

Points worth remembering:
A good fragile watermark scheme with recovery capabilities must generate s summary bits for n bits of principal content where $s \ll n$.

7.1.3 Summary and Authentication Bit Streams

Now we understand what summary and authentication bit streams are all about. A problem arises here where we have $k + s$ (where k is authentication and r is recovery bits) number of bits to generate, embed and extract. Because now we are interested in localizing the tampered region as well as recovering it also. The problems of $k + s$ bit generation, embedding and extraction can be further bifurcated into the pixel-based and block-based techniques. We will see all the problems and their solution in upcoming sections in detail.

Points worth remembering:
Now we have to generate, embed and extract $k + s$ bits where k is the number of authentication bits and s is the number of summary bits for n bits of principal content.

7.1.4 Fragility of Fragile Watermark

In the previous chapter, as we sawn that a fragile watermark plays the role of hash function for images, so it must follow all the properties of a hash function. A fragile watermark with recovery capabilities must also follow all the earlier mentioned properties of a fragile watermark. It will be destroyed even with a small change in the watermarked image. The basic difference in the fragility of fragile watermark with and without recovery capabilities is the embedding location. We have two bit streams to embed, one is for authentication information and called the authentication bit stream and the other one is a recovery bit stream called a summary bit stream. Now here the challenge is to embed the authentication/summary bit stream in such a way that it becomes more/less fragile in nature. Thus, after detection of the altered location we would be able to recover it. We will discuss the embedding procedure in detail in the next section.

Points worth remembering:
Fragility of the authentication/summary bit stream must be high/low as per the requirement at the receiver end.

7.1.5   Types of Fragile Watermark with Recovery Capabilities

Categorization of fragile watermarks with recovery capabilities can easily be understood from Figure 7.1. Here we can see that fragile watermarks with recovery capabilities can be broadly classified in two ways:

1. On the basis of the embedding mechanism.

2. On the basis of the extraction mechanism.

7.1.5.1   On the basis of the embedding mechanism

When a fragile watermark with recovery capabilities is generated, it must be embedded into the cover image in such way that it can effectively do its assigned job (tamper detection and recovery of tampered principal content). On the basis of fragile watermark embedding techniques, it is further classified into four categories:

1. **Fragile watermark with block-based authentication and block-based recovery :** As its name suggests, in this category, the authentication bit stream for integrity verification and the summary bit stream for recovery are generated for the blocks of the given cover image. First of all the cover image is divided into non-overlapping blocks of fixed size. After that, the generated watermark with both bit streams is embedded into that block or any mapped block. We have already seen that the main advantage of this type of fragile watermarking scheme is its low complexity because we do not need to process each and every pixel. We treat whole block as a single pixel. The block-based fragile watermarking scheme is less accurate because if a single pixel of the given block is altered, then the whole block will be treated as an altered block, in spite

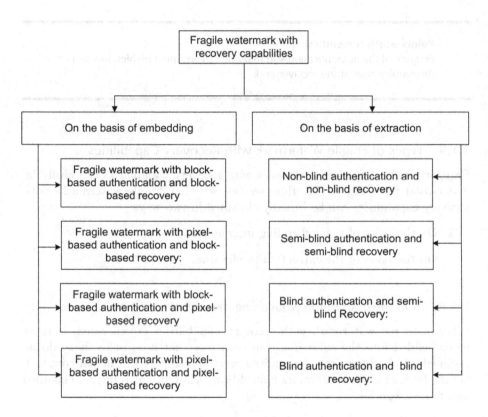

FIGURE 7.1 Classification of fragile watermarks with recovery capabilities.

of having a number of pixels unaltered. Hence this type of fragile watermark is unsuitable for exact localization of tampering. This scheme is useful when we need to know only whether the given watermarked image is authentic or not. Similarly when we perform recovery with this scheme, then a staircase problem occurs in the recovered image. To minimize this problem, we need to take blocks of very small size.

2. **Fragile watermark with pixel-based authentication and block-based recovery:** In this category, the authentication bit stream is generated using all pixels of the given cover image but the summary bit stream is generated the same as the previous one, on the basis of blocks. This scheme is most suitable for exact localization of tampering and approximate recovery of the tampered region. If the size of the cover image is very large, then the complexity of this algorithm is very high because authentication bits are generated on a pixel basis.

3. **Fragile watermark with block-based authentication and pixel-based recovery:** This category is treated as a hypothetical class of fragile watermark because it is neither practically feasible nor desired. In this category, the authentication bit stream is generated using a block of the given cover image but the summary bit stream is generated based on the pixel level. This scheme is infeasible because when we have less space to store the authentication bit stream k (due to block-based authentication), then how and where we will store the summary bit stream s as the length of the summary bit is greater than the length of the authentication bit stream is a problem. This scheme is less desired because in the case of block-based tamper localization due to false acceptance, we unnecessarily mark those pixels as altered which were actually not altered. That is why recovery will be done for the whole block which may reduce the imperceptibility between the watermarked and recovered image.

4. **Fragile watermark with pixel-based authentication and pixel-based recovery:** This category of fragile watermark is the ideal category. Here both authentication and summary bit streams are generated at the pixel level and only those pixels will be detected and recovered which are actually altered. In this case the imperceptibility between watermarked and recovered images will be more satisfactory. The only drawback of this approach is it is computationally more complex.

Points worth remembering:
Block-based fragile watermarks are less complex but without exact tamper localization.

Points worth remembering:
One can remove the stair case effect in the recovered image with block-based recovery by taking blocks of very small size.

Points worth remembering:
Fragile watermarks with pixel-based authentication and summary bit streams are ideal cases and more desirable in terms of imperceptibility between watermarked and recovered images but they are computationally more complex.

7.1.5.2 On the basis of the extraction mechanism

At the receiver end, once a watermarked image is reached, we need to extract the watermark in order to check the integrity of the watermarked image and if it is tampered with, then we need to recover it for any further applications. On the basis of watermark extraction, fragile watermarks may be further classified into the following categories:

1. **Non-blind authentication and non-blind recovery:** In this category of fragile watermark, we require the original cover image for tamper localization and its recovery at the receiver end. This class of watermark is highly undesirable because it increases the overhead for storing and transmitting the cover image along with the watermarked image. If authentication/recovery is non-blind, it means we are transmitting the original cover image also with the watermarked image. Hence in this case there is no meaning of semi-blind or blind recovery/authentication.

2. **Semi-blind authentication and semi-blind recovery:** In this category of fragile watermark, instead of the whole cover image, we require only some side and abstract information of the cover image at the receiver end, to detect the tampered region and its recovery. Sometimes we embed the authentication and recovery information of the principal content only, so at the receiver end we also need to pass the embedding locations for both bit streams. As we know that the length of the authentication bit stream is always less than the length of the summary bit stream. Hence if authentication is semi-blind (required some abstract information at the receiver end) then recovery cannot be blind, and we must need some mapping information for recovery. We need to transmit this side or abstract information over a secure channel or this information must be encrypted to prevent it from any middle attack.

3. **Blind authentication and semi-blind recovery:** In this category no information is required for tamper detection except the symmetric key with which the watermark was embedded. But some abstract information related to mapping of the summary bit stream is required for approximate or perfect recovery of the tampered region of the watermarked image.

4. **Blind authentication and blind recovery:** This category of fragile watermark is most desirable because it does not require any related information about the cover image at the receiver end during tamper detection or for the recovery process. Watermark embedding and extraction algorithms are well known to all, hence to prevent the unauthorized extraction or embedding of invalid watermarks into the cover image, we need some symmetric key known to only the sender and receiver. So in the case of blind authentication and recovery, we need only a symmetric key at the receiver end which was used during embedding of the watermark.

Points worth remembering:
There should be a secure mechanism to transmit the auxiliary information required for extraction of the watermark at the receiver end.

7.1.6 Ideal Category for Fragile Watermark with Recovery Capabilities

We have seen many types of fragile watermark with recovery capabilities classified on the basis of their embedding and extraction mechanisms. Now the question arises about which category may be called ideal and why. A fragile watermark will be more desirable if embedding of authentication and summary bit streams is done on a pixel wise basis and blind extraction of both bit streams is done at the receiver end, because in this category, we can perfectly localize the tampered locations as well as recover those locations with good imperceptibility. There will be no need of any auxiliary information related to the cover image or recovery mapping at the receiver end due to the blindness of the approach. Still an effective algorithm is missing in the literature for this type of approach.

Points worth remembering:
A fragile watermark will be more desirable if embedding of authentication and summary bit streams is done on a pixel wise basis and blind extraction of both bit streams is done at the receiver end.

7.2 GENERATION OF A FRAGILE WATERMARK

In the previous section we went through various categories of fragile watermarks with recovery capabilities. Now our objective is to generate a fragile watermark with recovery capabilities for a given cover image in order to secure and manage its integrity. In this section we will see various concepts which can be applied to generate an efficient fragile watermark. These approaches are called fragile watermarking techniques with recovery capabilities.

7.2.1 Self-Embedding Techniques

In Chapter 6, we saw the basic concepts for a self-embedding approach. In self-embedding approaches, authentication information is generated from the pixel itself and again embedded into the corresponding pixel. We do not need any additional image for watermarking. Similarly in the case of fragile watermarking with recovery capabilities, the recovery information is also generated either on the block basis or pixel basis along with authentication information and again embedded into the corresponding blocks or pixels. In this chapter we are only focused on recovery in the spatial domain, hence we will see only those approaches which operate directly in the spatial domain.

With self-embedding techniques, one can make one's own fragile watermarking approach. On the basis of authentication and recovery bit generation, self-embedding techniques can be understood in three ways because a fourth one is not possible:

1. **Self-embedding with block-based authentication and block-based recovery**

2. **Self-embedding with pixel-based authentication and block-based recovery**

3. **Self-embedding with pixel-based authentication and pixel-based recovery**

7.2.1.1 Self-embedding with block-based authentication and block-based recovery

Here our attention is concentrated on blocks rather than pixels. This type of procedure is used when we are more sensitive to the quality of the watermarked image and hence we have much less space for watermark (authentication and recovery bit streams) embedding in the cover image. The following procedure is applied for this category of self-embedding:

1. First of all the cover image I of size $M \times N$ is divided into non-overlapping blocks of size $m \times n$. Here $m << M$ and $n << N$.

2. For each block, we have to generate authentication bit stream k and summary bit stream s where $k << s$.

3. Now each block will be treated as a small image, and authentication bit stream k is generated by any of the block-based methods discussed in the previous chapter.

4. Since we have seen how to generate authentication bits for a block in the previous chapter, we are not focusing much on that topic. Here our attention is focused on the generation and embedding of summary bit stream s.

There are various ways in which we can make our own algorithm to generate a summary bit stream for an image block. Here we will see a few examples so that we can be familiar with the basic modus operandi of recovery bit generation for a cover image.

Let I be an image of size $M \times N$. A pixel of $(x, y)^{th}$ location in I is denoted by $P(x, y)$. If we are dealing with gray scale cover images, then a pixel can be represented as eight bit binary form and $P_b(x, y)$ represents the bth bit of the pixel where $b = 1$ to 8 (1 represents LSB and 8 represents MSB). Since we are dealing with the block-based techniques, image I will be divided into non-overlapping blocks of size $m \times n$. Now we need to generate authentication bit stream k for each block of I. Assume that we have generated k for each block of I using any given method in the previous chapter. Now our objective is to generate a summary bit s for each block.

Before doing anything related to the algorithm, there are some preliminary steps which must be followed :

1. First of all decide how many bits will be dedicated to embedding of each pixel in the block so that the watermarked image will be visually appealing and tamper detection can be performed effectively (see details in the previous chapter). Suppose we are taking u bits from each pixel in the block of size $m \times n$, then we have total $m \times n \times u$ number of bits per block for embedding of any information related to authentication or recovery.

2. We have already assumed that we have generated authentication bit stream k for each block. Then we need to allocate the space for the authentication bit stream from the total number of embedding space $m \times n \times u$ bits. Now the available storage for the summary bit stream is $m \times n \times u - k$.

3. Because we are taking u bits from each pixel of the block, we are less concerned about the original u LSBs of each pixel of the block. We know that these u bits of each pixel will be updated from the watermark information. Now the total number of bits of the principal content in each block for which we have to generate a summary bit stream is $(m \times n)(8 - u)$.

4. Now we have to generate $m \times n \times u - k$ number of summary bits for $(m \times n)(8 - u)$ number of the principal content's bits where $m \times n \times u - k << (m \times n)(8 - u)$.

5. Our objective is to generate a type of algorithm for summary bit generation which can return the desired number of bits that can perfectly or approximately recover the principal content of the block.

Example of summary bit generation for block-based recovery:

Here we provide some samples of summary bit generation approaches. The reader can generate his own better approach on the basis of these concepts.

1. **Summary bit generation based on the mean value of the block:**
   In this case we replace all the pixel values of the block with their mean value at the receiver end. It means that if we consider a block of small size then this approach will work fine, but in the case of large block size, the staircase effect can be seen at the receiver end. We can understand this approach from example 7.1:

   Example 7.1: Let us consider an image I of size $M \times N$. Since we are dealing with block-based authentication and recovery, the image will be divided into non-overlapping blocks. Assume that the size of the block is 2×2 for this example. Hence there will be four pixels in each block. Let $B = \begin{bmatrix} 156 & 155 \\ 150 & 153 \end{bmatrix}$ be a block of size 2×2 from image I. Here we are taking two LSBs from each pixel, hence $u = 2$. So we have a total $m \times n \times u$ i.e. $2 \times 2 \times 2 = 8$ bits in each block for embedding of authentication and recovery information of $m \times n(8 - u)$ i.e. $2 \times 2 \times 6 = 24$ bits. First of all we convert these pixels of the block into binary form, so the matrix will be like $B = \begin{bmatrix} 10011100 & 10011011 \\ 10010110 & 10011001 \end{bmatrix}$. Since we are taking $u = 2$ bits for each pixel, remove the first two LSBs from all pixels, and now the block will be $B = \begin{bmatrix} 100111 & 100110 \\ 100101 & 100110 \end{bmatrix}$ or $B = \begin{bmatrix} 39 & 38 \\ 37 & 38 \end{bmatrix}$. Next we calculate the mean of all these pixels $\frac{39+38+37+38}{4}$ i.e. 38. It means if we represent the watermark bit as W_b, then $W_b = s + k$ where s is the summary bit stream and k is the authentication bit stream. Till now we have calculated six bit summary bit stream $s = 100110$, but we still have space for two bits of authentication. Let us consider that in reference to the previous chapter, we have generated two authentication bits for this block using a block-based self-embedding approach and these bits are $k = 10$. Now append these two authentication bits with the summary bits in order to get W_b so $W_b = 10011010$ for the given block. We have eight bits for the watermark and two bits available in each pixel for embedding. Now the question arises of where we should embed these eight bits. We will see this problem in the next section but right now we are embedding these watermark bits W_b in the same originating block and the updated block will look like $B_w = \begin{bmatrix} 10011110 & 10011001 \\ 10010110 & 10011010 \end{bmatrix}$ or $B_w = \begin{bmatrix} 158 & 153 \\ 150 & 154 \end{bmatrix}$. We can see that there is not much difference between the original block B and the watermark embedded block B_w. Visually these pixels look similar, only we can quantitatively find out the difference between both blocks. One can understand the aforementioned example graphically using Figure 7.2. This example shows block-based recovery and authentication. The colored bits in W_b show the summary bits whereas the remaining bits show authentication bits.

2. **Summary bit generation based on any clustering method:** With the help of the above mentioned example or method, we understand our objective, in order to generate a summary bit stream. This objective can be achieved by many other methods. Summary bit generation using the mean method may fail when pixels of the block are not approximately similar. If the differences between all pairs of pixels are very high in the block, then replacement of all pixels by their mean at the receiver end, does not make sense. This is because the mean value will be completely different either in terms of visualization or PSNR. In such a scenario we can think about different methods. Suppose we take the block of size 4×4 so each block will contain 16 pixels. It is not necessary that all pixels will be of similar or approximately similar intensities. In such a scenario we can use different clustering methods like K-means, K-medoids to make the clusters of similar intensities. Since we have 16 pixels and if we are taking two bits from each pixel, then still we have $16 \times 2 = 32$ bits in one block for embedding the summary as well as authentication information. So here we also have to put the mapping cluster for each block. One can understand the concept of a cluster-based method by seeing Figure 7.3. Here the K-means clustering method is used in order to generate three clusters. We can see here that all pixel intensities which are in the same cluster are much less different and can be replaced by the cluster mean. The values in two clusters are sufficiently different. If we apply a mean-

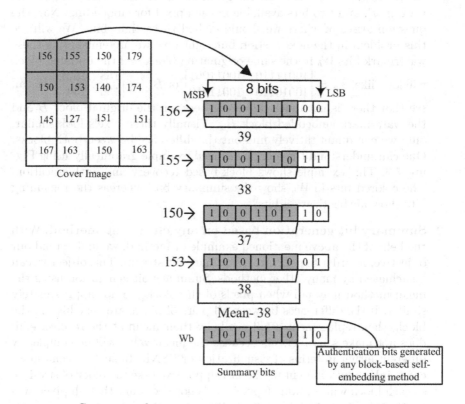

FIGURE 7.2 Summary bit generation using mean-based method.

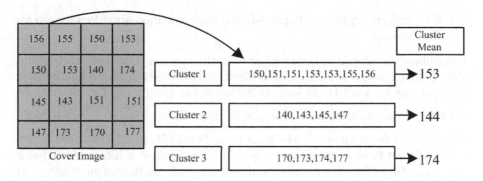

FIGURE 7.3 Summary bit generation using cluster-based method.

based method, in this example, then all pixels would be replaced by an overall single mean i.e. 157. Here there are many pixels which are very different from 157 hence the recovery image will not be pleasant enough.

Readers can pursue this direction in order to make their own algorithm for summary bit generation in cases of block-based recovery.

MATLAB Code 7.1

```
%MATLAB Code 7.1
%MATLAB code for block based processing.
cover_image=double(imread('lena.bmp'));
[r c]=size(cover_image);
m=2;
NofoBlock= (r*c)/(m*m); % Calculate the number of blocks
watermarked_image=cover_image;
x=1;
y=1;
for (kk = 1:NoofBlock)
block=cover_image(y:y+(m-1),x:x+(m-1));% Assigning the block of size m*m to variable block
 % Do any operation on block of size m*m
watermarked_image(y:y+(m-1),x:x+(m-1))= block; % Reassign the updated block to
     watermarked image.
   if (x+m) >= c
      x=1;
      y=y+m;
   else
      x=x+m;
   end
end
watermarked_image=uint8(watermarked_image);
figure
imshow(cover_image,[]);
figure
imshow(watermarked_image,[]);
```

7.2.1.2 Self-embedding with pixel-based authentication and block-based recovery

If the above mentioned scheme is clear to us then we can very easily understand this method. Suppose we are doing block-based recovery using a mean-based method, but we want to do authentication at the pixel level. Let us continue with the same aforementioned example. Here $B = \begin{bmatrix} 156 & 155 \\ 150 & 153 \end{bmatrix}$ is a block of size 2×2 from image I. We are taking two LSBs from each pixel hence $u = 2$. So we have in total $m \times n \times u$ i.e. $2 \times 2 \times 2 = 8$ bits in each block for embedding of authentication and recovery information of $m \times n(8 - u)$ i.e. $2 \times 2 \times 6 = 24$ bits. Since we are doing pixel-based authentication, we must reserve at least one bit for each pixel for authentication. For the given example, each block contains four pixels, so then we must reserve at least four bits for authentication. Suppose we have generated four authentication bits for the given four pixels using any pixel-based self-embedding method mentioned in the previous chapter and that is 1010. Now we have only $8 - 4 = 4$ bits for the summary in each block. Here we take the mean value by using six bits of each pixel in the block and the mean is 38. But we have only four bits to store this information, so we take its first four MSBs which is 1001. So the total watermark bit $W_b = 10011010$.

Here we have to pay attention to few things. Whenever we convert a decimal value into binary then it must be in eight bit format, for example, 35 intensity will be converted into eight bit binary, e.g., 00100011. If we calculate for MSBs in order to calculate the mean or anything, then we have to consider the appended zero also. For example if for embedding of W_b, we remove two LSBs, then the remaining MSBs will be 001000. In the aforementioned example, the summary bit stream 1001 will only get its actual weightage when we preserve its value with its bit position. It means 1001 will only get its actual meaning when recovery is performed by appending four 0000 at last i.e. 1001000 in order to preserve bit position. Now we can see that the decimal value of only 1001 is 9 but of 10010000 is 144, which is very near the intensities mentioned in the block. The same example can be understood graphically by Figure 7.4. Here the colored bits in W_b show the summary bits generated by the block-based mean method and the remaining four bits are authentication bits generated by the pixel-based method. In this example, one authentication bit is dedicated to one intensity value.

7.2.1.3 Self-embedding with pixel-based authentication and pixel-based recovery

This approach is an ideal case of fragile watermarking with recovery capabilities. Till now we have understood that this approach is ideal and also tough enough to implement. Since each pixel has eight bits of information, and if we assign $u = 4$ for watermark embedding of each pixel, then we have to accommodate the summary as well as the authentication information in four bits

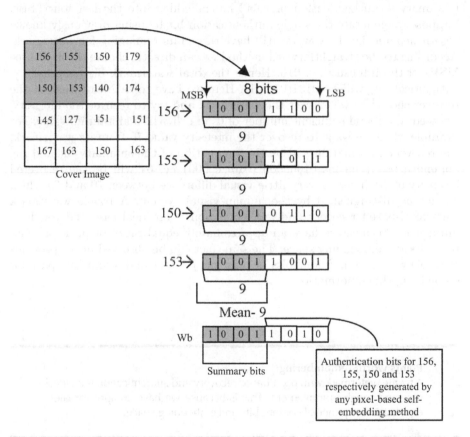

FIGURE 7.4 Fragile watermarking with block-based recovery and pixel-based authentication.

per pixel. In the last chapter, we saw that if we increase the number of bits per pixel for watermark embedding, then it will reduce the imperceptibility of the watermarked image. If we generate a single pixel authentication bit for each pixel then we still have $4 - 1 = 3$ bits for recovery information. A straightforward method is to use direct mapping of a pixel with its approximate pixel. For example if the intensity value is 75, then its eight bit binary value is 01001011. Now if $u = 4$ then it means watermark the information (summary + authentication bit) will be embedded into the last four LSBs. Suppose we generate the single authentication bit by using previously known algorithms and i.e. 1. Now we still have three bits remaining for watermark. According to the straightforward method we can directly choose the first three MSBs of the intensity i.e. 010. Hence the total watermark bit (summary + authentication) will be 0101(010+1). Here we have to take care that at the receiver end when we extract the watermark and try to recover the intensity, we need to append sufficient number of 0s in order to maintain its value. For example when we want to recover the intensity value 70 from its watermark, the recovery information will be the summation of the summary bit and the appended 0s. It means 01000000 (010+00000) i.e. 64 will be the recovered intensity of 70. There is very little visual difference between 70 and 64 which cannot be distinguished by the human visual system. A fragile watermark with pixel-based recovery and authentication is an ideal case and tough to implement, because we have an approximately equal ratio for principal content bits and embedding space. The same fact can be observed in the previous example where we had four bits of principal content and a four bit space for embedding that information.

Points worth remembering:
A fragile watermark with pixel-based recovery and authentication is an ideal case and tough to implement. This is because we have an approximately equal ratio for principal content bits and embedding space.

One can make an efficient fragile watermarking approach for pixel-based recovery and authentication by using some compression techniques in order to reduce the size of the principal content bits. For example if we apply the run length encoding method to reduce the size of the principal content bit, then the ratio of principal content bits and embedding space will not be the same. Once we have a fewer number of principal content bits then we require a lower number of bits for embedding, and in this way we can maintain the imperceptibility between the watermarked image and cover image.

Points worth remembering:
An efficient fragile watermarking approach with pixel-based recovery and authentication can be developed using some bit compression mechanisms in order to change the ratio of embedding space and principal content bits.

7.3 EMBEDDING OF FRAGILE WATERMARK

In the previous section we saw various block-based and pixel-based mechanisms for summary and authentication bit generation. Now our next step in watermarking is embedding of a fragile watermark. In the case of a fragile watermark with recovery capabilities, embedding is little bit tricky compared to the approaches mentioned in the previous chapter. Here the block for which the watermark information is generated is called the **Originating Block** and the block in which the watermark will be embedded is called **Embedding Block**. We can categorize the embedding process on the basis of originating and embedding blocks, into three categories:

7.3.1 Embedding and Originating Blocks are Same:

This category is least desired. In this case we embed the watermark information of a block into the same block. This approach was suitable when we only required authentication but now we also want recovery of the tampered principal content. If the recovery information will be embedded into the same originating block, then in the case of tampering, it may be possible that the summary bit would be destroyed. In this case we would not be able to recover the tampered principal content.

7.3.2 Originating Blocks are Embedded into Sequentially Mapped Embedding Blocks

In this category we do not embed the watermark information into the same mapping block in order to preserve the summary bit in spite of having an attack. Here we use a secret number which set the start of mapping of embedding and originating blocks. Let us assume that we are using a block of size $m \times n$ in the image I of size $M \times N$, so the total number of blocks will be $\frac{M \times N}{m \times n}$. Suppose the starting mapping secret number is n, which means that the ith originating block will be mapped into $\{(n+i) \bmod \frac{M \times N}{m \times n}\}^{th}$ embedding block. The drawback of this type of embedding is the secret mapping number n because this is the bottleneck of this approach. If an attacker were to know this

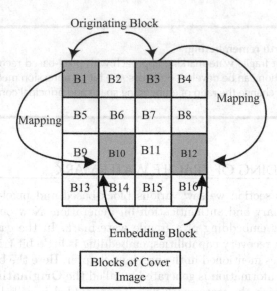

FIGURE 7.5 Sequential mapping of originating and embedding blocks.

number then he could easily extract the watermark information and change accordingly. Figure 7.5 shows the sequential mapping of originating and embedding blocks. In this example $n = 9$ which is why block 1 is mapped with $(1+9) \bmod 16 = 10^{th}$ block and block 3 is mapped with $(3+9) \bmod 16 = 12^{th}$ block.

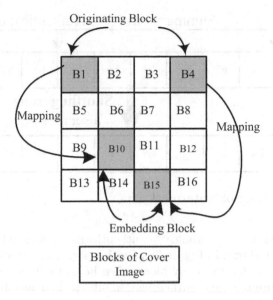

FIGURE 7.6 Random mapping of originating and embedding blocks.

MATLAB Code 7.2

```
%MATLAB Code 7.2
%MATLAB code for generating the index for sequential mapping.
cover_image=double(imread('lena.bmp'));
[r c]=size(cover_image);
m=2;
n=2;
NofoBlock= (r*c)/(m*n); % Calculate the number of blocks
Sec= input('Enter the value of secret key for mapping');
for i=1:NoofBlock
    map(i)= mod((sec + i), NoofBlock);
end
%array map will contain the mapping index of the originating and embedding
%blocks
```

7.3.3 Originating Blocks are Embedded into Randomly mapped Embedding Blocks

In order to improve the previous approach we can map the originating block with any randomly selected embedding block. Here the problem is that we need to keep track of mapping blocks for recovering or authenticating the tampered blocks. We also take care that the watermark information of more than one originating blocks must not be mapped with a single embedding block. Otherwise we can lose the information for the other blocks except the last one. Figure 7.6 shows the random allocation of originating blocks into the embedding blocks.

One can enhance the security of the embedding process by randomly shuf-

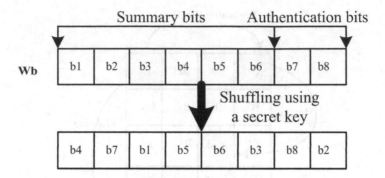

FIGURE 7.7 Shuffling of watermark bits.

fling the watermark bits (summary + authentication) of each block by using a symmetric key as shown in Figure 7.7. By this way if an attacker extracts the watermark information from any blocks then he will still be unable to distinguish between summary and authentication bits as these are shuffled. We need to reshuffle these bits using the same symmetric key and only then we will be able to detect the tampering and recover the tampered content. Embedding is done at the pixel level as mentioned in previous chapter. Since still we want to localize the tampered pixels or blocks, we need good fragility in the watermark which is why all watermark information is embedded into the spatial domain of the image. The selection of the number of bit planes is still a crucial issue for watermarked image quality. Since we now have more watermark bits due to summary bits, imperceptibility of the watermarked image will definitely be compromised. Our objective is to a such type of fragile watermarking approach with recovery capabilities which supports good imperceptibility with perfect tamper localization and recovery.

Points worth remembering:
The security of watermark embedding can be enhanced by shuffling the watermark bits (summary + authentication) using a symmetric key.

7.4 EXTRACTION OF A FRAGILE WATERMARK

This phase is executed at the receiver end. This phase is very important because it deals with tamper localization as well as recovery of tampered principal content. An image may be tampered with either at the time of transmission or at the time of storing that image. We have already studied in the previous chapter that tampering may be either intentional or unintentional. In the introductory section we saw that on the basis of watermark extraction, a watermarking approach is further classified into various categories. In the case of fragile watermarking with recovery capabilities, the output of the watermark extraction phase will be the location of the tampered pixels as well as their recovered version.

7.4.1 Tamper Localization and Recovery

Tamper localization and recovery are done on the basis of types of fragile watermark mentioned in the previous section. A fragile watermark is categorized into four classes on the basis of watermark extraction mechanisms. Tamper localization and recovery for each category are a little bit different.

7.4.1.1 Non-blind authentication and non-blind recovery

As we know that in the case of non-blind fragile watermarks, we need an untampered version of the watermarked image at the receiver end for tamper localization as well as recovery. Pixel wise comparison is done to localize the tampering. Sometimes there is some tampering which cannot be identified just by the human visual system, and for that case we require pixel wise comparison of untampered and tampered watermarked images in this category. If any tampering is found, we simply replace the tampered values with untampered versions.

7.4.1.2 Semi-blind authentication and semi-blind recovery

In this category of tamper localization and recovery, we do not require an original watermarked image at the receiver end. But here we need some side or abstract information of the watermarked image which contains the locations for watermark embedding. These locations contains all the mapping information which is required at the time of tamper localization and recovery. Once we identify the locations where a watermark is embedded then we simply extract the watermark from the LSBs. Then we again generate the authentication bits of the fragile watermark using same the algorithm by which it was generated at the time of embedding. Now we have two sets of authentication bits, one is the extracted one and the other one is recalculated one. We simply do pixel wise comparison on these bits and if any mismatch is found between the recalculated and extracted ones then that pixel or block is marked as tampered. Now with the help of the same mapping information, we again extract the summary bits for the recovery of tampered pixels or blocks.

7.4.1.3 Blind authentication and semi-blind recovery

In this class, we do not require any side or abstract information of mapping for tamper localization except a symmetric key. We only extract and recalculate the authentication bits for the entire set of pixels or blocks of the image. If any mismatch is found, then we need to recover it again and for that recovery we need some side or abstract information about the mapping blocks or pixels which contains the summary bits.

7.4.1.4 Blind authentication and blind recovery

In this class, we do not require any side or abstract information about mapping for tamper localization or recovery of the tampered principal content except a symmetric key. We only extract and recalculate the authentication bits for the entire set of pixels or blocks of the image. If any mismatch is found, then we need to recover it again and for that recovery use some symmetric key to find out the mapping blocks or pixel. Once we reach this point, there then just extract the summary bits and recover the tampered information again.

Tamper detection and its recovery result is affected by the means of generating the fragile watermark. As we know that watermark can be embedded either block wise or pixel wise. Hence we can find the accuracy accordingly.

7.4.1.5 Block-based tamper detection and recovery

If a fragile watermark (summary + authentication bits) is generated by the block-based self-embedding technique then we cannot perform exact localization of the alteration in images as well as we cannot achieve exact recovery also. In this case, image is divided into non-overlapping blocks and cumulative authentication bits and the summary bits are generated using all pixels of that block. These generated fragile watermark bits are embedded into the parent block or any other mapped block. If any intentional or unintentional tampering is done to any pixel of that block, the entire block will be treated as altered. In Figure 7.8, the image of Lena is watermarked using block-based authentication and the block-based recovery approach and some alteration is performed on the watermarked image in Figure 7.8(b). But we can see in Figure 7.8 (c) that the whole block is treated as tampered though it also contains those pixels, which are not actually tampered. Figure 7.8 (d) shows the recovered image. Since we are using block-based recovery, we can see that the recovered version of the tampered portion of the watermarked image is not perfect. It has less PSNR value quantitatively even though human visual system is able to identify the actual content of the original watermarked image.

7.4.1.6 Pixel-based tamper detection and block-based recovery

If the authentication bits of the fragile watermark are generated using a pixel-based self-embedding method whereas the summary bits are generated using a block-based self-embedding method, then we can exactly localize the tampered region but cannot perfectly recover it again. In this case at least one authentication bit is dedicated to each pixel of the block but cumulative re-

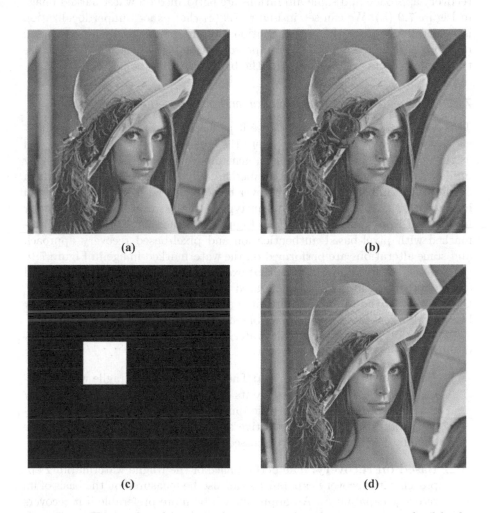

FIGURE 7.8 Example of tamper localization and its recovery for block-based authentication and recovery approach: (a) Watermarked image,(b) Tampered image, (c) Tamper localization, (d) Recovery of the tampered portion.

covery information is generated. Both of these bit streams are then embedded into the originating block or any other mapping block. In Figure 7.9, the image of Lena is watermarked with pixel-based authentication and a block-based recovery approach and some alterations are performed on watermarked image in Figure 7.9 (b). We can see in Figure 7.9 (c) that exact tamper localization is performed because the authentication bits are generated by a pixel-based method. But the recovery of the tampered portion is still not effective because the summary bits are generated by the block-based method.

7.4.1.7 Pixel-based tamper detection and pixel-based recovery

This category is most desired because it perfectly localizes the tampered pixels as well as recovers them also. Here the authentication bit stream as well as the summary bit stream are both generated on a pixel level. That is why unlike the block-based method, this method only localizes those pixels which are actually tampered with and recover those pixels with good accuracy and imperceptibility. The perfection of this type of fragile watermark can be better understood from the Figure 7.10. In Figure 7.10, the image of Lena is watermarked with pixel-based authentication and pixel-based recovery approach and some alterations are performed on the watermarked image in Figure 7.10 (b). We can see in Figure 7.10 (c) that exact tamper localization is performed as the authentication bits are generated by a pixel-based method. Similarly, the recovery of the tampered portion is also very effective because summary bits are also generated by the pixel-based method.

The performance of the fragile watermarking approach with recovery capabilities can be judged in three ways:

1. **Based on tamper detection:** The efficiency of the fragile watermarking approach can be judged by its tamper detection rate. As we know that a pixel-based authentication method is appropriate for exact tamper localization. The quantitative analysis of tamper detection can be done by FAR and FRR as discussed in the previous chapter.

2. **Based on recovery:** The performance of the fragile watermarking approach with recovery capabilities can also be measured on the basis of its recovery capabilities. An approach will be more preferable if it recovers the tampered portion of the image with good imperceptibility. That is why pixel-based recovery is preferable.

3. **Based on imperceptibility between watermarked and cover images:** The fragile watermark will be composed of two types of information: one is the summary bit stream for recovery and the other one is authentication bit stream for tamper detection. So we have a huge amount of data to be embedded into the cover image. Now we have to take care of the quality of the watermarked image after embedding. Embedding of a fragile watermark must be very effective so that the im-

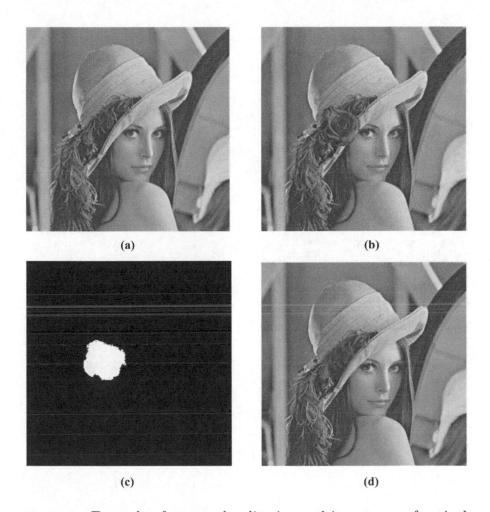

FIGURE 7.9 Example of tamper localization and its recovery for pixel-based authentication and block-based recovery approach: (a) Watermarked image,(b) Tampered image, (c) Tamper localization, (d) Recovery of the tampered portion.

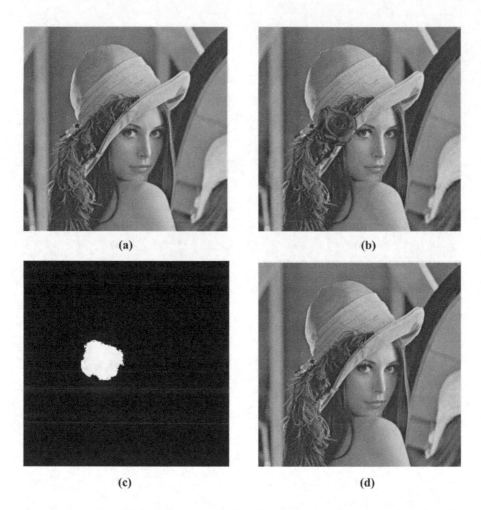

FIGURE 7.10 Example of tamper localization and its recovery for the pixel-based authentication and pixel-based recovery approach: (a) Watermarked image,(b) Tampered image, (c) Tamper localization, (d) Recovery of the tampered portion.

perceptibility between the cover and watermarked image has good range of PSNR.

7.4.2 Non-Zero FAR Due to Improper Mapping

In the last chapter we saw that all approaches will have zero value for FAR because we cannot treat any unaltered pixel as altered. This is done only because of the proper mapping of the originating and embedding pixels or blocks. If the originating and embedding pixels or blocks will be the same then FAR will definitely be zero but this scenario is not possible in case of recovery. Here we have to hide the recovery information for any pixel or block into another place so that after destroying those corresponding pixels or blocks, we will be able to recover them again. It means that if the originating pixels or blocks will not be the same as the embedding ones then we can suffer from the problem of non zero FAR. The reason behind it is, an attack may alter all of the MSBs and LSBs of the given pixel. LSBs contain the mapping fragile watermark of another block or pixels whereas the authentication and summary bit for their MSBs are stored somewhere else. If the pixel (including its MSBs and LSBs) is altered, then due to a change in MSBs that will be treated as altered and due to change in LSBs, the mapping pixel or block will be treated as altered, which is not actually altered. That is why we suffer from the problem of non-zero FAR. To remove this problem, proper mapping is required for originating and embedding blocks or pixels.

Points worth remembering:
Improper mapping of originating and embedding blocks or pixels causes non-zero FAR(False Acceptance Rate).

SUMMARY

- *A fragile watermark with recovery capabilities* can be used to assist in verifying the integrity and recovering of the cover work.

- A fragile watermark with recovery capabilities can be broadly classified in two ways: *on the basis of embedding and on the basis of extraction.*

- On the basis of embedding, fragile watermarking can further be classified in four ways: *block-based authentication and block-based recovery, pixel-based authentication and block-based recovery, block-based authentication and pixel-based recovery, pixel-based authentication and pixel-based recovery.*

- On the basis of extraction, fragile watermarking is further classified in four ways: *non-blind authentication and non-blind recovery, semi-blind authentication and semi-blind recovery, blind authentication and semi-blind recovery, blind authentication and blind recovery.*

- A *self-embedding approach* for fragile watermark generation is the best way to generate our own algorithms.

- A fragile watermark with recovery capabilities is the combination of two bit streams: *authentication and summary bit streams.*

- The self-embedding approach for this type of watermark is classified in three ways: *block-based authentication and block-based recovery, pixel-based authentication and block-based recovery, pixel-based authentication and pixel-based recovery.*

- The *spatial domain* of an image can be used for the best processing of the fragile watermarks with recovery capabilities.

- *LSBs* are the most suitable place for fragile watermark embedding.

- *Tamper localization and recovery* is done on the basis of type of the fragile watermark.

- Tamper localization includes the process of comparison between extracted and recalculated fragile watermarks.

- The efficiency of fragile watermarking with recovery capabilities can be judged by its *False Rejection Rate (FRR)* and *False Acceptance Rate (FAR)*, Imperceptibility and quality of recovered portion of the image and imperceptibility between cover and watermarked images.

- Three types of mapping procedures are available for originating and embedding blocks and pixels.

- Improper mapping of originating and embedding blocks or pixels causes non zero FAR.

Fragile Watermark with Recovery Capabilities in Frequency Domain

CONTENTS

I N the previous chapter, we saw how we can use a fragile watermark with recovery capabilities in the spatial domain in order to authenticate and to recover the tampered version of the image. We have already seen that the spatial domain is more vulnerable to any intentional or unintentional attack. That is why tamper localization can easily be done in this domain but due to the high fragility of the watermark (which is necessary as well) recovery information may also be lost. Hence we have two challenges, one is to make a fragile watermark with high fragility so that we can localize even a small change in pixel values of the watermarked image and second, we have to preserve the recovery information i.e. the summary bit in spite of any attack. To achieve this objective, we can generate authentication bits in the spatial domain and summary bits using frequency domain. As we know that the frequency domain is more robust against intentional or unintentional attacks so it is more suitable for summary bit generation.

Hence in this chapter we deal with various theoretical and practical aspects of fragile watermarks with recovery capabilities in the frequency domain and this concerns the authentication and restoration of the associated cover images. The **Chapter Learning Outcomes(CLO)** of this chapter are given below. After reading this chapter, readers will be able to:

1. Understand different techniques used for recovery in the frequency domain.

2. Compare pros and cons of different technologies in fragile watermarking with recovery capabilities.

3. Write their own fragile watermarking algorithms with recovery capabilities in the frequency domain.

4. Identify the research gap and research interest in this field.

5. Implement their own scheme in MATLAB.

8.1 INTRODUCTION

We have seen many theoretical and practical aspects regarding fragile watermarking schemes. We have seen how the spatial domain plays the most important role in fragility. A fragile watermarking scheme with recovery capabilities in the spatial domain is capable enough of recovering the tampered region of the image approximately, but this will only happen when we are able to preserve the summary bits even after an attack. As we know that the spatial domain is more vulnerable to attacks because we operate directly on the pixel values, hence it is good enough for tamper localization. Since recovery information is also generated directly from the pixel values themselves, if the summary bit stream is modified, then we cannot recover it anyhow. We must have some mechanism so that if the summary bit stream is modified up to a certain extent then we still must be able to recover the approximate value of the tampered location. This will only be possible when we generate a summary bit stream in the frequency domain. Since in the frequency domain, each coefficient is calculated with the help of all the pixel values of the image, they are highly dependent on each other. In case of tampering, we can recover at least the nearest values of the frequency coefficients from remaining original pixel values of the watermarked image. Each domain has its own pros and cons in order to make a fragile watermark equipped with recovery capabilities. In this chapter we deal with the fragile watermark with recovery capabilities in the frequency domain. In upcoming sections, we will see the benefits and drawbacks of working in the frequency domain.

8.1.1 Summary Bit Stream: Recovery Information through Frequency Coefficient

In approaches shown in the previous chapter, the source of the summary bit stream was the pixel intensities as well as the embedding location which was also the spatial domain itself. In the case of the frequency domain, recovery information will be generated by the frequency coefficients but it can be embedded into either the spatial domain or the frequency domain. Here all notations related to authentication and summary bits will be the same, only the generation and embedding methods will be changed.

Points worth remembering:
In fragile watermark schemes with recovery capabilities in the frequency domain, summary bits are generated by using the frequency coefficients of the image.

8.1.2 Types of Fragile Watermark with Recovery Capabilities in Frequency Domain

As we know that authentication bit streams will always be generated by using pixel intensities and always be embedded into the spatial domain itself, because of its high vulnerability to attack. This condition is not compulsory in case of the summary bit stream. Summary bit streams can be generated using pixel intensities and embedded into the spatial domain as we have seen in the previous chapter. Similarly, summary bit streams can also be generated by frequency coefficients and they can be embedded into the spatial domain or into frequency domain. In the previous chapter we saw the authentication and summary bit generation using pixel intensities and embedding into the spatial domain, hence in this chapter, categorization of fragile watermarks

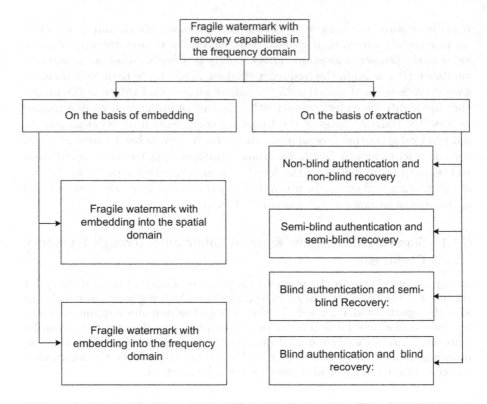

FIGURE 8.1 Classification of fragile watermark recovery capabilities in frequency domain.

with recovery capabilities in the frequency domain will only be based on the generation and embedding of the summary bit stream. Here we can stick with the generation process of a summary bit stream into the frequency domain only because we have already dealt with summary bit generation in the spatial domain in the previous chapter. Classification can easily be understood from Figure 8.1. Here we can see that fragile watermarks with recovery capabilities in the frequency domain can be broadly classified in two ways:

1. On the basis of the embedding mechanism.

2. On the basis of the extraction mechanism.

8.1.2.1 On the basis of the embedding mechanism

When a fragile watermark with recovery capabilities in the frequency domain is generated, it can be embedded either in the spatial domain or in the frequency domain. Hence we can further categorize it into two types on the basis of the embedding domain:

1. **Embedding into the spatial domain :** In this category we assume that authentication bits are generated using pixel intensities themselves and the summary bit stream is generated in the frequency domain but embedded into the spatial domain. In this category, due to intentional or unintentional attacks, it may be possible that along with authentication bits, summary bits are destroyed, as both are embedded into the spatial domain. Since summary bits are generated in the frequency domain, hence in spite of destroying the few or more summary bits, we can still recover the approximate information of the principal content of the tampered location.

2. **Embedding into the frequency domain:** In this category we assume that authentication bits are generated by pixel intensities themselves and the summary bit stream is generated as well as embedded into the frequency domain. Since here we do not directly embed the summary bits into the spatial domain, in this approach, summary bits are less vulnerable to attacks. Here we have to take care of the embedding sequence for authentication and summary bits, and we will see this in more detail in the next section.

8.1.2.2   On the basis of the extraction mechanism

In the previous chapter, we categorized the fragile watermark with recovery capabilities on the basis of the extraction mechanism into four categories: Non-blind authentication and non-blind recovery, semi-blind authentication and semi-blind recovery, blind authentication and semi-blind recovery and blind authentication and blind recovery. When we deal with the fragile watermark with recovery capabilities in the frequency domain, then the same categorization with their actual meaning can be followed.

8.2   GENERATION OF A FRAGILE WATERMARK

In the previous chapter we saw various approaches to generate a fragile watermark in spatial domain. We saw that the self-embedding approach is the best way to generate authentication and summary bit streams. In this section we see how the self-embedding approach can be used to generate a summary bit stream in the frequency domain. Although our watermark bits consist of two bit streams, namely, summary and authentication bit streams, in this chapter we are ignoring the generation mechanism for authentication bit streams which we have already seen in the previous chapter. Here we discuss the concepts of summary bit generation in the frequency domain so this may be one more reason to ignore the generation of an authentication bit stream.

8.2.1 Self-Embedding Techniques

In earlier chapters, we saw the basic concepts of the self-embedding approach. In the self-embedding approach, authentication as well as recovery information in generated from the pixel or any feature itself and again embedded into the corresponding pixel. A summary bit stream may be generated for a single pixel or for a block having more than one pixel. In case of recovery in the frequency domain we do not directly deal with the pixel value or any block of the pixels. First of all we change its domain using any frequency transformation and after that we generate the recovery information using the corresponding co-ordinate's coefficient. In Chapters 6 and 7, we understood that authentication and summary bit streams are both independent entities and an authentication bit stream can be generated using any pixel or block-based method irrespective of the generation of a summary bit stream. The generated authentication bits are just appended with the summary bit stream in order to give it the complete watermark information. That is why in this section we categorize the self-embedding approach only on the basis of a summary bit stream due to the simple fact that authentication bit streams can only be generated in the spatial domain in either ways (pixel or block).

On the basis of only summary bit generation with the frequency domain's coefficients, self-embedding techniques can be understood in two ways:

1. **Block-based self-embedding approach**

2. **Non-block-based self-embedding approach**

8.2.1.1 Block-based self-embedding approach

Unlike a block-based approach in the spatial domain, here recovery of tampered blocks may be done pixel by pixel within a block. Here a block is only used to reduce the complexity of summary bit generation because, as we know, each frequency coefficient is calculated using all pixel intensities of the image. Hence we divide the image into non-overlapping blocks of size $m \times n$ and each block is treated as an image. We apply any transformation block wise, not on the whole image as shown in Figure 8.2. We get better results of recovery when working in the frequency domain because in spite of using blocks of an image, we can do recovery on a pixel basis which is not possible in cases of block-based recovery in the spatial domain. The followings are the basic steps for performing recovery in the frequency domain.

1. Let I be a cover image of size $M \times N$ and w be the watermark bits for I which consist of two bit streams, namely, authentication and summary bit streams denoted by k and s, respectively.

2. Generate authentication bit stream k using any block or pixel-based approach in the spatial domain just by using pixel intensities of the cover image I.

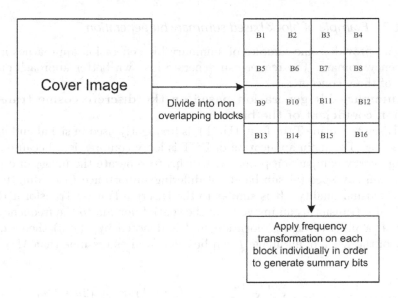

FIGURE 8.2 Mechanism for the block-based self-embedding approach.

3. In order to generate a summary bit stream in the frequency domain, we apply a transformation like the Discrete Cosine Transformation (DCT), Discrete Wavelet Transformation (DWT) etc based on its applicability and characteristics.

4. Now we have the matrix of transformation coefficients I_{TC} having same size $M \times N$ as the original cover image I.

5. Divide the matrix I_{TC} into non-overlapping blocks of size $m \times n$. Here $m << M$ and $n << N$.

6. For each block, we need to generate a summary bit stream s where $k << s$, hence we apply our a self-embedding approach on each block.

7. Since we have seen how to generate authentication bit for a block in previous chapter hence we are not focusing much on that topic. Here our attention is focused on the generation of summary bit stream s with the help of frequency coefficients.

There are various ways in which we can create our own algorithm to generate a summary bit stream for a frequency coefficient block. Here we will see a few examples so that we will be familiar with the basic modus operandi of recovery bit generation for a cover image in the frequency domain.

8.2.1.2   Example of block-based summary bit generation

Here we provide some samples of summary bit generation approaches in the frequency domain. The reader can generate his own better approach on the basis of these concepts.

Summary bit generation based on the discrete cosine transformation coefficient of the block:

The Discrete Cosine Transform (DCT), is frequently used in signal and image processing. The main application of DCT is lossy compression, because of its strong energy compaction property. It helps to separate the image into parts or we can say spectral sub-bands of differing importance (according to the image's visual quality). It is similar to the Discrete Fourier Transform(DFT) which also transforms an image from the spatial domain to the frequency domain. If a pixel of $(x, y)^{th}$ location in I is denoted by $f(x, y)$ then $(u, v)^{th}$ entry of the DCT of an image I can be calculated as (assume that $M = N$)

$$F_{(u,v)} = C(u)C(v) \sum_{x=0}^{N-1} \sum_{y=0}^{N-1} f_{(x,y)} \cos \frac{(2x+1)u\pi}{2N} \cos \frac{(2y+1)v\pi}{2N}, \qquad (8.1)$$

where $0 \leq u, v < 8$ and

$$C(u) = \begin{cases} \sqrt{\frac{1}{N}} & u = 0 \\ \sqrt{\frac{2}{N}} & u > 0 \end{cases}. \qquad (8.2)$$

Since $M = N$, the value for $C(u)$ and $C(v)$ will be the same. Here $u, v = 0, 1, 2...N - 1$. Based on the calculated DCT coefficients, we proceed further in order to calculate the summary bit stream. There may be many methods based on the DCT coefficients but here we are suggesting one which may be helpful to generate the reader's own method.

Step 1. Divide the input cover image I into non-overlapping blocks of size 2×2, hence the counting of the blocks is $\frac{N^2}{4}$ and the ith block of image I is denoted by I_b^i where i is from 1 to $\frac{N^2}{4}$. Now remove the first three LSBs of each pixel intensity of the blocks. The resulting image is called content image and the ith block of content image is denoted by I_c^i.

Step 2. Since each pixel intensity of an image is of five bits now, pixel of the content image will have thirty-two possible values ranging from 0 to 31. To center (normalize) it on zero, subtract each pixel by half the number of maximum possible values, i.e. sixteen.

$$I_c^i = I_c^i - 16, i = 1, 2, 3, ..., N^2/4 \qquad (8.3)$$

Step 3. Now we apply a block wise DCT operation on each block of content image I_c. The resultant block having the transformation coefficients is denoted by I_{TC}.

$$I_{TC}^i = DCT(I_c^i), i = 1, 2, 3, ..., N^2/4 \qquad (8.4)$$

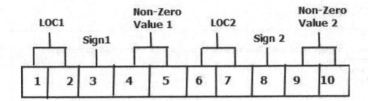

FIGURE 8.3 Assignment of summary bits in a vector w.

Step 4. Do point wise division of each element of the DCT coefficient matrix with reduced upper left quantization matrix q of size 2×2 that is taken from a common JPEG recommended quantization matrix as shown below. Through rigorous observation we have concluded that at least two DCT coefficients will always be zero in this case.

$$SparseMatrix(R^i) = round(\frac{I^i_{TC}}{q}), i = 1, 2, 3, ..., N^2/4 \qquad (8.5)$$

where quantization matrix

$$q = \begin{bmatrix} 16 & 11 \\ 12 & 12 \end{bmatrix} \qquad (8.6)$$

Step 5. Insert the non zero values of R^i with its sign into vector w as shown in Figure 8.3.

$$sign(x) = \begin{cases} 0 & \text{if } x > 0; \\ 1 & \text{if } x < 0. \end{cases} \qquad (8.7)$$

where x is a non-zero value.

Step 6. Now we have to preserve the exact location of the non zero values in R^i so that at the receiver end we can easily recover it. This work is done using the following location matrix

$$M_L = \begin{bmatrix} 00 & 01 \\ 10 & 11 \end{bmatrix} \qquad (8.8)$$

Step 7. Now we will put the ten summary bits in following manner in vector w as shown in Figure 8.3.

MATLAB Code 8.1

```
%MATLAB Code 8.1
% This MATLAB code is for generation of Sparse matrix for above mentioned block based
      summary bit generation.
cover_image=double(imread('lena.bmp'));
[r c]=size(cover_image);
m=2;
n=2;
Quantization_Matrix= [16 11;12 12];
NofofBlock= (r*c)/(m*n); % Calculate the number of blocks
Sparse_Matrix=zeros(r,c);
x=1;
y=1;
for (kk = 1:NoofBlock)
block=cover_image(y:y+(m*n-1),x:x+(m*n-1));% Assigning the block of size m*n to variable
      block
Normallized_block= block-16;
Transformed_block= dct2(Normallized_block) ;
S_M= round(Transformed_block .\ Quantization_Matrix);
Sparse_Matrix(y:y+(m*n-1),x:x+(m*n-1))= S_M; % Assign the 2x2 sparse matrix.
    if (x+m*n) >= c
       x=1;
       y=y+m*n;
    else
       x=x+m*n;
    end
end
```

Here the spatial locations of the first and second non zero value in the matrix R^i are denoted by $Loc1$ and $Loc2$. Again by rigorous observation it is concluded that the non zero values always range from -2 to 2 hence we require only two bits to represent the absolute values for it.

Let us consider that we are taking three bit LSBs from each pixel of the block for watermark embedding, so we have in total $4 \times 3 = 12$ bits for watermark embedding. In the above method we have generated 10 bits with the help of frequency coefficients of a block of size 2×2 which holds the summary information of that block. Now we have two more bits left for embedding of authentication information. Since we have four pixels in each block it means that we cannot generate pixel wise authentication bits as we have only two bits for authentication. Here two bits are generated using block wise authentication and these two bits are appended into the watermark vector w with summary bits. Now this generated watermark vector will be embedded into the spatial domain of the image. In this case tamper detection and recovery of the tampered pixel will be a little bit difficult because we need to reprocess all the steps of generating summary bits in the reverse manner on the received bits to get the DCT coefficients. Once we get the coefficient, we apply inverse DCT to get the actual pixel intensity on that place.

Once we locate the tampered block by using two authentication bits we need to recover that block. So for restoring the block, steps are as follows.

Step 1. First of all initialize a 2×2 empty matrix M_e and fill the corresponding two non zero locations as mentioned in vector w and fill zero at the remaining two locations.

Step 2. Apply element-wise multiplication to the matrix M_e with quantization matrix q, and find the inverse-DCT of that in the matrix r having size 2×2.

$$r = Inverse - DCT(M_e \times q) \qquad (8.9)$$

Here inverse DCT can be calculated in the following way,

$$f_{(x,y)} = \sum_{u=0}^{N-1} \sum_{v=0}^{N-1} C(u)C(v)F_{(u,v)} \cos \frac{(2x+1)u\pi}{2N} \cos \frac{(2y+1)v\pi}{2N}, \qquad (8.10)$$

where $x, y = 0, 1, 2..N - 1$.

Step 3. Calculate the floor value of each element of the matrix r by adding 16 and normalize the corresponding value.

$$r_{ij} = \lfloor r_{ij} + 16 \rfloor \qquad (8.11)$$

$$r_{ij} = \begin{cases} 0 & \text{if } r_{ij} < 0 ; \\ r_{ij} & \text{if } 0 \le r_{ij} \le 31 ; \\ 31 & \text{if } x > 31. \end{cases} \qquad (8.12)$$

Step 4. Convert the r_{ij} in five bits and insert it into the locations of the five MSBs of the corresponding pixels in the tampered block of the watermarked image. Now we have recovered pixels in five bit form, and in order to convert it into the eight bits, just append three 0s in the first three LSB's location so that each gray value ranges from 0 to 255. So we finally recover the tampered block in a pixel-by-pixel manner with approximate intensities.

We can say that in the above approach we are doing authentication in a block-based manner whereas recovery in a pixel-based manner which looked impractical in the case of recovery in the spatial domain (chapter 7). Actually in the above method, we compact the principal content of the block very effectively and precisely which is not possible in the spatial domain. In spite of embedding these generated watermark bits into the spatial domain these bits can still tolerate little bit of an attack which was completely impossible in recovery in the spatial domain. Suppose summary bits are changed due to an attack, since each coefficient is calculated by all other elements of the block, we can still recover the approximation of the pixel. But in case of recovery in the spatial domain, summary bits are embedded directly into the LSBs of the pixels, so once these are destroyed due to an attack, we cannot recover the tampered pixel again.

Points worth remembering:
In case of recovery in the frequency domain, pixel-based recovery with block-based authentication is possible which is difficult in the case of recovery in the spatial domain.

Points worth remembering:
We can generate more compact summary bits in the case of recovery in the frequency domain.

Points worth remembering:
Summary bits generated by the frequency domain are more robust to attacks as their approximation can be generated in spite of an attack.

Till now we have seen a DCT based method for tampered region recovery which is based on block wise computation of frequency coefficients. This is not the only method. We have shown just an approach so that the reader can think in that direction and create his own efficient method.

8.2.1.3 Non-block-based self-embedding approach

Readers may think that in the previous chapter there were two classifications of self-embedding approach, namely, block-based and pixel-based, but in this chapter, we have replaced the pixel-based approach with a non-block-based approach. The reason behind this is the passive contribution of pixel intensities in frequency domain. Actually in the frequency domain operation, we do not deal directly with the pixel intensities, we deal with the frequency coefficient of that coordinate. We can also compute the cumulative coefficients on any coordinate value based on its neighbour's coefficients. That is why we call it a non-block-based method in place of a pixel-based method. In the previous section we were focused on a block-based operation on the frequency coefficients of the image. Now we will discuss those techniques in which we do not divide the transformed image into non-overlapping blocks. That is why these methods can be called a non-block-based self-embedding approach. Here we directly apply recovery techniques on the coefficients. We can see in Figure 8.4 how frequency transformation is applied on the overall image as a whole without dividing it into blocks. Many methods can be developed to complete the task of recovery. To create an efficient method we need to understand the property of every frequency transformation including DCT, DWT, DFT etc and choose the appropriate one.

8.2.1.4 Example of summary bit generation for non-block-based recovery

Here we provide some samples of non-block-based self-embedding approaches for summary bit generation in the frequency domain. the reader can generate his own better approach on the basis of these concepts.

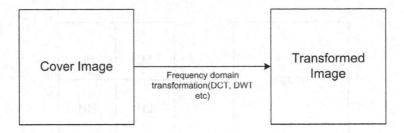

FIGURE 8.4 Mechanism for a non-block-based self-embedding approach.

Points worth remembering:
Self-embedding approaches can be categorized as block-based and non-block-based self-embedding approaches.

Points worth remembering:
Since we use frequency coefficients in place of pixels for self-embedding, that is why we call this a non-block-based self-embedding approach.

Summary bit generation based on the discrete wavelet transformation coefficient of the cover image:
Discrete wavelet transformation converts the cover image into a finite series of wavelets that can be stored more efficiently than spatial pixel blocks. Wavelets have a sufficient number of rough edges and they are perfectly able to render the image. An image can be transformed by using 2-Dimensional wavelet transform generally known as 2-D DWT. 2-D DWT can be observed as a 1-D wavelet scheme which transforms along the rows and then a 1-D wavelet transform along the columns, respectively. The 2-D DWT is applied in a very straightforward manner by just inserting an array transposition between the two 1-D DWTs. First of all the rows of the array are processed with only one level of decomposition. This essentially splits the array into two vertical halves, with the first half storing the average coefficients, while the second vertical half stores the detail coefficients. The same process is repeated further with the columns, resulting in four sub-bands as shown in Figure 8.5 within the array defined by the filter output. Figure 8.6 (d), (e) and (f) show a one-level, two-level and three-level 2-D DWT of the Lena image, respectively.

The aforementioned figures show how discrete wavelet transformation can reduce the size of the principal content of the cover image into the one fourth dimension. This transformation highly compresses the cover image and we can easily generate summary bits from its reduced version of principal contents.

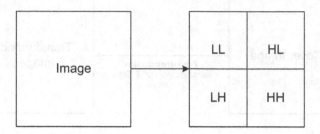

FIGURE 8.5 Representation of four sub-bands of an image in one-level DWT.

FIGURE 8.6 Four sub-bands of one-level, two-level and three-level DWT.

Since in this method, we directly apply the transformation on the whole image without making any block or division, this approach can be called a non-block-based self-embedding approach.

8.3 EMBEDDING OF A FRAGILE WATERMARK

In the previous section we saw various block-based and non-block-based mechanisms for generation of summary bits. Our next step in watermarking is embedding of these summary bits. In case of recovery in the frequency domain, we make the assumption that we always embed the authentication bit stream into the spatial domain because the spatial domain is more vulnerable to attacks and the job of authentication bits is to locate the tampering due to any attack. Now the main challenge is to decide the embedding domain for summary bits. We can embed the summary bits either in the spatial domain or in the frequency domain. So based on the embedding domain, we can categorize the fragile watermarking with recovery in the frequency domain into two categories.

8.3.1 Embedding of Summary Bits in Spatial Domain

In this case we embed the summary bits of a block into the spatial domain of the image as shown in Figure 8.7. In this figure we can see that we apply transformation on each block separately and generate summary bits for each block. Once we generate the summary bits, we embed them into the same or any other mapping block in the spatial domain of the image. Since authentication bits are already embedded into the spatial domain, we can trace any tampering done via intentional or unintentional attacks. Summary bits are generated through the frequency domain which is why all summary bits are tightly coupled. When the watermarked image is altered, then due to embedding of summary bits into the spatial domain, the summary bit may be altered. Since frequency coefficients are highly dependent on each other, so we can generate approximation of the tampered region even after attack. To avoid false acceptance of the untampered region as being tampered, we need to map the summary and authentication bits of a block into another block very effectively as discussed in the previous chapter.

Points worth remembering:
Frequency coefficients are highly coupled to each other which is why any tampered value can be recovered approximately with the help of other frequency coefficients of the image.

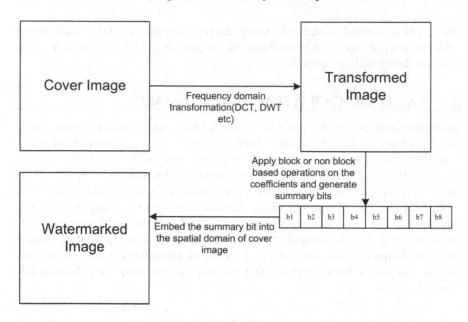

FIGURE 8.7 Embedding of summary bits into the spatial domain of the cover image.

8.3.2 Embedding of Summary Bits in Frequency Domain

In this category we do not embed the summary bits into the spatial domain of the image in order to preserve the summary bit in spite of having an attack. Here the calculated summary bits are embedded into the frequency domain of the image. Now the question arises of how to embed the summary bits into the frequency domain. Actually in this case we modify the frequency coefficients according to the corresponding summary bit's value as shown in Figure 8.8. Suppose we have generated four binary summary bits (like mean value of the block) for four pixels of the block of size 2×2. Then each bit will correspond to a single pixel of the block. Since we have to embed these summary bits into the frequency domain, we will modify the frequency coefficients accordingly as per the value of 0 and 1. For example if we have to embed the summary bit value 0,then we can make the corresponding frequency coefficient value even, otherwise make it odd. We can choose any other combination or methods so that we can identify the actual summary bit values as 0 or 1 at the receiver end. As we are embedding the summary bits in the frequency domain, they will be robust for most of the attacks.

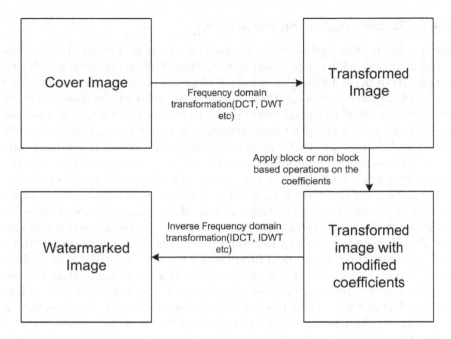

FIGURE 8.8 Embedding of summary bits in frequency domain.

Points worth remembering:
Embedding of summary bits into the frequency domain means modifying the frequency coefficients in such a way that one can easily identify the embedded bits (either 0 or 1).

8.4 EXTRACTION OF A FRAGILE WATERMARK

As we already know that this phase deals with tamper localization and its recovery but in the case of recovery in the frequency domain, we require one more piece of information regarding the embedding domain. We have two objectives at the receiver end, one is tamper localization and the other one is recovery. In the previous chapter we saw that on the basis of information needed for tamper detection and recovery, we can categorize fragile watermark in four ways, namely, non-blind authentication and non-blind recovery, semi-blind authentication and semi-blind recovery, blind authentication and semi-blind recovery and non-blind authentication and non-blind recovery. All of these classes of watermarking approaches have the same characteristics and properties as discussed in Chapter 7.

8.4.1 Tamper Localization and Recovery

Tamper localization and recovery are done on the basis of types of fragile watermark mentioned in the previous section. First of all we extract the authentication bits from the LSBs of the pixels. If authentication bits are generated for the blocks of the cover image, then tamper localization will be done in a block wise manner. Similarly pixel wise tamper localization is done, if pixel-based authentication bits are generated. Summary bits may be embedded either in the spatial domain or the frequency domain of the image. If summary bits are embedded into the spatial domain, then we extract summary bits for the tampered pixels or blocks from the corresponding mapped pixels or blocks. After getting the summary bits, we apply the reverse method in order to regenerate the approximate pixel/block value of the tampered one. Figure 8.9 shows an example of tamper detection and recovery with generation and embedding of summary bits in the frequency and spatial domains, respectively. Here (a) show the watermarked image which is tampered with up to 30% by a direct content removal attack shown in (b). Due to authentication bits we can see that we detected most of the tampered region shown by the white color in Figure 8.9(c). With the help of summary bits which are generated in the frequency domain, we can see that all tampered content is recovered very efficiently.

If summary bits are embedded into the frequency domain by modifying the frequency coefficients of the image, then first we need to calculate the required frequency coefficients of the image. Using the recalculated frequency coefficients, we again remap the approximate pixel values of the tampered region.

FIGURE 8.9 Example of tamper detection and recovery with generation and embedding of summary bits in the frequency and spatial domains, respectively.

SUMMARY

- *Summary bits generated in the frequency domain* are more robust against various types of intentional and unintentional attacks because of the high coupling of frequency coefficients.

- *In fragile watermarking with recovery capabilities in the frequency domain* , summary bits may be embedded either in the spatial domain or the frequency domain.

- On the basis of embedding, fragile watermarking can further be classified in two ways: *fragile watermark with embedding in spatial domain and fragile watermark with embedding in frequency domain.*

- On the basis of extraction, fragile watermarking is further classified in four ways: *non-blind authentication and non-blind recovery, semi-blind authentication and semi-blind recovery, blind authentication and semi-blind recovery, blind authentication and blind recovery.*

- A self-embedding approach may be either *block-based or non-block-based*

- A fragile watermark with recovery capabilities in the frequency domain is the combination of two bit streams: *authentication and summary bit streams*

- Authentication bits are always embedded into the spatial domain because it is more vulnerable to attack.

- *LSBs* are the most suitable place for fragile watermark embedding.

- Computational complexity is increased due to frequency transformation of the image at the sender and receiver end.

- In the frequency domain we can generate *more compact summary bits.*

- Fragile watermarking with recovery capabilities in the frequency domain may perform *block-based authentication with pixel-based recovery* which seems to be impossible in the spatial domain because of the larger summary bit stream.

Robust Watermarking

CONTENTS

I N the previous chapter, we saw various basic and advanced concepts related to fragile watermarking approaches which can be utilized in order to analyse and to implement an efficient fragile watermarking scheme. Fragile watermarks can be used to assist in maintaining and verifying the integrity of the associated cover image. When any tampering is done in watermarked image, we can extract the watermark and localize the altered locations. In advance variants of fragile watermarks, we can recover the tampered principal content as well, without the help of the original cover image. In this chapter we will see a different kind of watermark which is actually used for very different applications. Some times we need to ensure our ownership of the cover image in case of a conflict. A fragile watermark is unsuitable for this type of application because it can be better utilized for integrity verification of the cover images. In order to ensure the ownership of the cover image, it must be equipped with some information regarding its owner. This information may be any copyright logo related to the owner of the cover image. Unlike fragile watermarks, this type of watermark must be sufficiently robust against any number of intentional or unintentional attacks. Because of its tolerance property against attacks, this type of watermark is called a robust watermark.

In this chapter we deal with various theoretical and practical aspects of robust watermarks which deal with the ownership of the associated cover images. The **Chapter Learning Outcomes(CLO)** of this chapter are given below. After reading this chapter, readers will be able to:

1. Understand the background and need for robust watermark.

2. Understand the different terminologies related to robust watermark.

3. Write their own robust watermarking algorithms.

4. Identify the research gap and research interest in this field.

5. Implement their own scheme in MATLAB.

9.1 INTRODUCTION

In the case of robust watermarks, fragility is highly undesirable because a robust watermark must survive in an environment of a number of intentional or unintentional attacks. Robustness is essential in order to check the ownership of the given cover image. The watermark is embedded into the cover image in such a way that if the content of the watermarked image is tampered with in any way, the associated robust watermark with that content will remain as it is. At the receiver end, the existence of the watermark confirms the ownership of the associated owner of the cover image. Hence our requirement is to make a type of robust watermark that is highly resistant to any kind of intentional or unintentional attack. Though this objective is very difficult to achieve, we will still try to make such robust watermarking techniques where the watermark can survive in an environment of maximized tampering.

9.1.1 Robust Watermark for Ownership Assertion

Why is robust watermarking required at all? We can understand this by one example. Suppose we are maintaining a website which deals with the intellectual property of the tribes of various regions. In this website we upload pictures of their traditional products which can be researched. After a few months, we see that the images we uploaded on our website are available on competitor's website. This is the time of conflict because we have to prove that those images are our property and stolen by that competitor. This problem can be easily solved if all our images are embedded with a robust watermark. We can use our organization's logo as copyright information and this information will be treated a watermark. This copyright logo will be embedded into the cover images using robust watermarking techniques so that if an attacker changes the watermarked image significantly, we can still extract the copyright logo and prove our ownership of those images. So first we change all the cover images which are going to be uploaded on the website into watermarked images. All the images will be embedded with the copyright logo of the website owner using robust watermarking techniques. If we see those images elsewhere, by simply extracting our watermark, we can confirm our ownership of those images. Robust watermarking is used to prove authorization.

Points worth remembering:
Robust watermarks are used for ownership assertion and to prove authorization.

9.1.2 Robustness of Robust Watermark

Why is this type of watermark called a robust watermark? In which context is a watermark robust? These types of various questions come in mind. Actually this type of watermark is called robust because it has the ability to resist various types of intentional or unintentional attacks. There are many routine types of images processing like noise removal, compression, rotation, resizing etc which are some times desired for a particular application. Hence our robust watermark must be sustainable in spite of having all these image processing operations. Actually sometimes these image processing operations are not for intentional removal of a watermark but for enhancement of the quality of the watermarked image. For this reason our watermarking approach must tolerate all these unintentional attacks as well as a satisfactory number of intentional attacks. Intentional attacks may include deliberately cropping a major portion of principal content of the watermarked image. There is no ideal algorithm present in the literature which can tolerate all kinds and amount of attacks. But our objective is to make an algorithm that maximizes the robustness against such type of attacks as much as possible.

Points worth remembering:
Robust watermarks are more resistant to various types of intentional and unintentional attacks.

9.1.3 Image as Watermark

In the case of fragile watermarks, self-embedding is the best approach by generating itself instead of taking a separate image as watermark. In a fragile watermark we need to create a dependent and a tightly coupled relationship between the cover image and watermark so that if any changes are made to the watermarked image, it must be reflected on the watermark also. Robust watermarks are entirely different from fragile watermarks because in this case we need to select a type of watermark which is highly unrelated and independent from the cover image. It means self embedding techniques are highly undesirable in cases of robust watermarking. It is challenging and peculiar task to generate watermark that is unrelated and independent from the cover image, and so we use a separate image as watermark. We need to create a watermarking algorithm which embeds this image watermark into the cover image so that robustness is achieved.

Points worth remembering:
A self-embedding approach is highly unsuitable for robust watermarking which is why we use any image as the watermark that is related to the owner of the cover image.

9.1.4 Relationship between the Dimensions of Cover and Watermarked Images

In order to maintain robustness, we need to embed a watermark image in such a way that in spite of having an attack on the watermarked image, we can extract the healthy watermark. To achieve this objective we need to embed a single intensity value of the watermark image into multiple locations of the cover image. Or, we could say that any one intensity of the watermark image is saved with the help of multiple intensities of the cover image. Thus, in case of any tampering, we can extract as much of the healthy watermark image as possible from the remaining (unaltered) locations or intensities of the watermarked image. This scenario shows that the dimensions of the watermark image must be smaller than the dimensions of the cover image. Let the dimension of the cover image be $M \times N$ and of the watermark image be $m \times n$ where $m \times n << M \times N$. We can conclude that in the case of a robust watermark, there will not be two types of categories, e.g., pixel-based and block-based as in fragile watermarks. Here everything is block-based because we need to embed a single piece of information of the watermark image into multiple locations or intensities or we could say in a block of the cover image.

Points worth remembering:
The Dimension of the watermark image will always be smaller than the dimensions of the cover image.

Points worth remembering:
Single intensity information of the watermark is embedded into multiple locations of the cover image.

9.1.5 Image Types for Watermark

In previous sections we saw that the relationship between the dimension of the watermark and cover image is $m \times n << M \times N$. This means that it will be better to minimize the bit information per intensity of the watermark image so that it can be saved properly into multiple locations. We have three options for image type which can be used for the watermark: first is the binary watermark, the second is a gray scale watermark and the last one is a color watermark. In the case of a binary watermark we require only a single bit per watermark intensity. Similarly in the case of gray scale and color images we need 8 bits and 24 bits information per intensity, respectively. We can conclude here that as per the requirement of minimizing the bit information of the watermark image, binary images are well suited for watermarks as they require a minimum number of bits $m \times n \times 1$ to represent the watermark of size $m \times n$ whereas gray scale and color images require $m \times n \times 8$ and $m \times n \times 24$ bits, respectively. That is why most of the existing approaches for robust watermarks are based on binary watermark images.

Points worth remembering:
Binary images are the most suitable type of watermark that can be used in robust watermarking as they take very few bits to represent the principal content of the watermark.

9.1.6 Single vs. Multiple Robust Watermarks

Sometimes embedding of a single watermark into the cover image may be undesirable. When the amount of tampering is huge, at that time it may be possible that all or a major portion of the watermark may be destroyed. In such scenario, it will be very difficult to prove ownership without the help of copyright information. That is why the concept of embedding of multiple watermarks into a single cover image came into the picture. These multiple watermarks may be identical or different copyright information related to the owner. Multiple watermarks may be embedded either with the same embedding approach or each watermark with a different embedding approach. Multiple watermarkings will be feasible if we deal with binary types of watermarks. Binary watermarks take very fewer number of bits in comparison to gray and color watermarks.

Points worth remembering:
Multiple watermarks can be used in order to improve the efficiency for verifying ownership.

9.1.7 Encrypted Robust Watermark

There are many situations when we need to embed an encrypted robust watermark instead of the original robust watermark. We can understand this form by one simple example. Let us consider that we have an organization and on its website, we uploaded all watermarked images. The copyright logo of this organization is used as a watermark. Since the copyright logo of any organization is well known, it may be possible that any attacker can embed our copyright logo into any objectionable image in order to defame us. To avoid this situation we need to embed the encrypted copyright logo instead of the original copyright logo. Encryption will be done using a secret key so that only the owner of the organization can decrypt it. The attacker cannot generate the same encrypted watermark unless he knows the secret key. By this way we can avoid this type of conflict and at the time in emergency, simply by using the same secret key, we can extract and decrypt the watermark in order to prove ownership.

Points worth remembering:
Encrypted watermarks can be embedded in order to avoid malicious embedding of authentic watermarks into the objectionable image.

9.1.8 Types of Robust Watermark on the Basis of Extraction Mechanism

Unlike fragile watermarking approaches, we cannot classify the robust watermarking approaches on the basis of the embedding mechanism, because we know that pixel-based embedding is impractical in this case. Since the size of the watermark is smaller than the cover image, each intensity of the watermark can be preserved in multiple locations or intensities of the cover image. Hence we can say that in robust watermarking, default embedding is block-based. At the receiver end, once the watermarked image is received, we need to extract the watermark in order to check the ownership of the received image. Categorization of robust watermarks can easily be understood by Figure 9.1. Here we can see that robust watermark can be broadly classified into three types on the basis of the extraction mechanism:

9.1.8.1 Non-blind robust watermark

In this category of robust watermark, we require the original cover/watermark for ownership identification at the receiver end. This class of watermark is highly undesirable because it increases the overhead for storing and transmitting the cover as well as the watermark image along with the watermarked one.

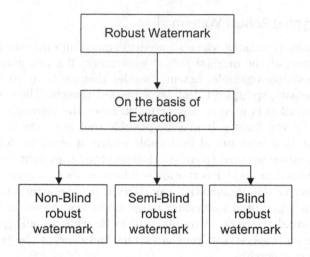

FIGURE 9.1 Classification of robust watermarks.

9.1.8.2 Semi-blind robust watermark

In this category of robust watermark, instead of the whole cover/watermark image, we require only some side and abstract information of the cover/watermark image at the receiver end, for ownership assertion. Sometimes we embed the watermark in some region-of-interest (ROI) of the cover image (see in detail in the next section). In such case we need to keep track of the embedding locations of the watermark in the cover image. This track record can be treated as side or abstract information of the cover image and it is required at the receiver end in order to claim ownership. We need to transmit this side information over a secure channel or this information must be encrypted to prevent it from any middle attack.

9.1.8.3 Blind robust watermark

This category of robust watermark is most desirable because it does not require any information related to cover/watermark image at the receiver end during the ownership assertion process. As we know that watermark embedding and extraction algorithms are well known to all, hence to prevent the unauthorized extraction or embedding of invalid watermarks into the cover image, we need some symmetric key known by only sender and receiver. So in the case of blind robust watermarks, we need only a symmetric key at the receiver end which has been used during embedding of the watermark.

Points worth remembering:
There should be a secure mechanism to transmit the auxiliary information required for extraction of the watermark at the receiver end.

9.2 GENERATION OF ROBUST WATERMARK

In previous sections we already saw that self-embedding is not a suitable approach for robust watermarking. Here we use a watermark image that is either binary or gray scale or color. So, this phase of watermark generation is very simple in cases of robust watermarking. Robustness will be achieved through this phase of watermark embedding which we will see in the next section. In the watermark generation process, we need to convert the watermark image into a one-dimensional encrypted binary pattern (vector). This pattern will be treated as input in next the embedding phase.

9.2.1 Binary Vector Generation for Binary Images

There are various ways in which we can store the bit information of the watermark image into a vector. Binary images may be treated as the subset of the Z^2 integer space. Here one Z is used to represent x coordinates and another Z is used to represent the y coordinate. As we know that binary images have only two intensities either 0 (Black) or 1 (White). So we need to store only the locations of either 1 or 0. If in the Z^2 set, we store only the $(x, y)^{th}$ coordinate of all those intensities having value $1/0$, then it means that at all remaining locations will have intensities $0/1$. Suppose a binary vector is represented as V, then first we store the dimension of the watermark image into its binary form. Let the dimension of the cover image be $M \times N$ and the watermark be $m \times n$, then vector V will be initiated as $V = [Binary\ of(m), Binary\ of(n)]$. In order to distinguish the binary of the dimensions, we will fix the length of binary stream of m and n with the upper bound. Now we will see which intensity (0 and 1) has the minimum count in the image. Suppose intensity 0 has a much lower count compared to 1. In this scenario we store the binary stream of the coordinate values of only 0 intensities. Now the vector V will be updated as $V = [Binary\ of(m), Binary\ of(n), Binary\ of\ x\ coordinate\ of\ first\ 0, Binary\ of\ y\ coordinate\ of\ first\ 0]$. This vector V will be updated with the next coordinate value of the 0 and so on.

Sometimes our watermark image may be too small so that vector V generation by the above method is costly. In that case we just append all the rows/columns of the binary watermark image and store them in vector V. It means $V = [Binary\ of(m), Binary\ of(n), rowwise/columnwise\ appended\ binary\ watermark]$. This technique is more simpler than previous one.

9.2.2 Binary Vector Generation for Gray/Color Images

We already know that bit requirements to represent the gray/color images are larger than the binary. Actually gray scale images belong to the Z^3 integer space. Here the first two $Z's$ are used to represent coordinate values of the pixel intensities and the third Z is used to represents the intensity itself. A single gray scale intensity can be denoted by eight bit binary form, similarly a color intensity can be represented by twenty-four bits binary form where each set of eight bits is for red, green and blue, respectively. In this case vector V will be initialized by the dimension of the watermark into binary form and after that, appended by a binary bit set of gray or color intensities of the watermark image from row or column wise.

9.2.3 Encryption of the Binary Vector of Watermark

In previous sections we saw why we need to encrypt the watermark before embedding. Encryption is done with a symmetric key. It means the same key will be required at both ends for encryption of the binary pattern as well as for decryption of the binary pattern. Since vector V consists of only two values, 0s and 1s, encryption means just toggling the bits from 0 to 1 or from 1 to 0. Encryption can also be performed just by shuffling the bits. Any process of shuffling or toggling or any randomization on a binary bit pattern is done with a seed value so that at the receiver end, using the same seed value, we can perform the reverse process. This seed value is nothing but the symmetric key.

$$V_e = encryption(V, SecretKey) \qquad (9.1)$$

9.3 EMBEDDING OF ROBUST WATERMARKS

In order to get high robustness, we need to develop a good embedding mechanism for the watermark. In upcoming sections we will deal with the process related to embedding and extraction of robust watermarks, hence all schemes which deal with these processes can be treated as robust watermarking schemes. There are various factors like imperceptibility between watermarked and cover images, robustness etc must be kept in mind while making any good robust watermarking scheme.

9.3.1 Domain Selection for Embedding

We have already seen that there are two types of domains viz The spatial domain and the frequency domain, which can be used for embedding of watermark. We need to choose a domain which is perfect for robust watermark embedding for valid reasons. As we know that a good robust watermark can be judged by its level of robustness against various minor and major attacks. Hence we have to choose the domain for embedding which is highly resistant against the smallest to largest distortion of images. In the case of the frequency domain, transformation coefficients at a particular location of the pixel are calculated by all pixels of the images. That is why most of the transformations are reversible in spite of missing some principal content from the images due to major attacks. Hence we can say that the frequency domain is a kind of robust domain which can tolerate minor and major attacks and hence is highly suitable for robust watermarking. Pixel intensities are more vulnerable than transformation coefficients to any minor distortion. Even a small addition of noise can change the original values of the pixels, hence we can conclude that the spatial domain is more fragile and sensitive to minor distortion of images (as it directly deals with pixels) and so the spatial domain is not suitable for robust watermarking.

Points worth remembering:
For the embedding of the robust watermark, the frequency domain is more suitable than the spatial domain.

9.3.2   Embedding of Robust Watermark in Frequency Domain

Till now, we have seen that the frequency domain is the better way for embedding the watermark so that it becomes robust in nature. There are various transformations which are reversible in nature like Discrete Fourier Transformation (DFT), Discrete Cosine Transformation (DCT), Discrete Wavelet Transformation (DWT) etc. All of these transformations have a unique property and they are reversible which means in spite of having changes in frequency coefficients, we can still recover the approximate intensity values of the original image. We know that each frequency coefficient is calculated by the contribution of all intensities in the image, hence significant changes in coefficients can be recovered. We can understand the embedding process in the frequency domain by using some examples.

Points worth remembering:
There are various frequency domain transformations available like DFT, DCT, DWT etc which can be utilized for watermark embedding in robust watermarking approaches.

9.3.2.1 DCT-based embedding for robust watermarks

Here we assume that we have two inputs, one is cover image I in which we want to embed the watermark and the second one is the encrypted vector V of the watermark. Suppose for this particular example, the size of the cover image is $M \times N$ and the size of the encrypted watermark vector is $1 \times \frac{M \times N}{64}$. We already knew that embedding of robust watermarks is by default block-based. So, in this example, we can say that we can save one bit of vector V in the 64 intensities of the cover image. Or we can say that we can take non overlapping blocks of size 8×8 in order to embed bits of V. Here we mention one of the possible approaches for watermark embedding. Readers can create their own better approach based on this. We can follow these steps to embed the watermark:

1. Take the DCT of cover image I. Values of the DCT coefficients may be either negative or positive.

2. Divide the matrix into non-overlapping blocks of size 8×8.

3. Let each element of the block be labelled with 1 to 64 numbers.

4. Again, subdivide the blocks into four equal regions. Each region will contain sixteen DCT coefficient values.

5. Now we have $\frac{M \times N}{64}$ number of bits for the watermark as well as the same number of blocks in the cover image. So each block will store a single binary watermark bit.

6. We will pick three sub-blocks ($2^{nd}, 3^{rd}$ and 4th) of a block in order to store the single binary bit.

7. Pick the coefficients of the index 22/50/54 and 15/43/47 from the $2^{nd}/3^{rd}/4^{th}$ sub-blocks.

8. Calculate the mean of the absolute values of the coefficients of indexes 22, 15 then 50, 43 and 54, 47 as shown in Figure 9.2. The mean can be calculated in the following ways:

$$m_1 = \frac{|DCT(22)| + |DCT(15)|}{2} \tag{9.2}$$

$$m_2 = \frac{|DCT(50)| + |DCT(43)|}{2} \tag{9.3}$$

$$m_3 = \frac{|DCT(54)| + |DCT(47)|}{2} \tag{9.4}$$

Here $DCT(x)$ represents the DCT coefficient at the xth index value in the block.

9. If we are to embed 0, then we will increase the even indexed coefficients (22, 50, 54) by their corresponding mean values, else increase the odd index's coefficient (15, 43, 47) by their corresponding means. We can see this process in Figures 9.3 and 9.4 where we are increasing the even index coefficient's value in order to embed 0 and we are increasing the odd index coefficient's value in order to embed 1. Since coefficient's values may be either negative or positive this is why we use + and - both signs in the figures. We have to apply whatever operation will be appropriate there to maintain the difference between the odd and even index's coefficients equal or more than the corresponding mean values.

10. It means for watermark bit 0, we need to modify the DCT coefficients such that

$$DCT(22) > DCT(15) \& DCT(22) - DCT(15) \geq m_1 \qquad (9.5)$$

$$DCT(50) > DCT(43) \& DCT(50) - DCT(43) \geq m_2 \qquad (9.6)$$

$$DCT(54) > DCT(47) \& DCT(54) - DCT(47) \geq m_3 \qquad (9.7)$$

And for watermark bit 1, we need to modify the DCT coefficients such that

$$DCT(15) > DCT(22) \& DCT(15) - DCT(22) \geq m_1 \qquad (9.8)$$

$$DCT(43) > DCT(50) \& DCT(43) - DCT(50) \geq m_2 \qquad (9.9)$$

$$DCT(47) > DCT(54) \& DCT(47) - DCT(54) \geq m_3 \qquad (9.10)$$

11. After doing this operation for all the bits of watermark V, just take the inverse DCT. Thus, the image achieved will be treated as a watermarked image.

In the above method, we simply embed the encrypted watermark vector into the frequency domain of the cover image. The aforementioned method is a blind approach to watermarking because at the receiver end, we do not need the original cover image or watermark, and we do not even need any abstract information related to embedding. Now we will see how to extract the watermark. Suppose during transmission or an other reason, some tampering is done to the watermarked image. At the receiver end, we will decide the watermark bit value on a majority basis.

1. Compute the DCT of the tampered/untampered watermarked image.

2. Again, divide the matrix into non overlapping blocks of size 8×8 and numbered indexes from 1 to 64.

3. Now compare the coefficients of the indexes 22 with 15, 50 with 43 and 54 with 47. If there are two or more coefficients of the even indexes which are greater than the coefficients of the odd indexes, then we extract the bit 0 from that block, else extract 1 from that block as a watermark.

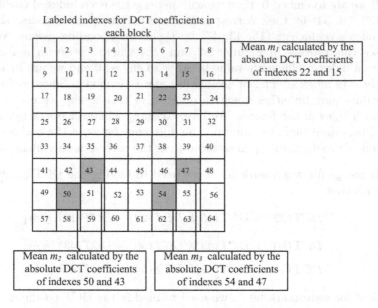

FIGURE 9.2 Calculation of mean for selected DCT coefficients.

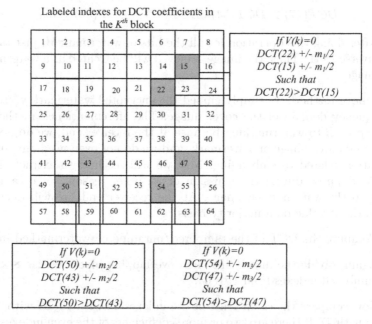

FIGURE 9.3 Watermark bit embedding process for vector bit 0.

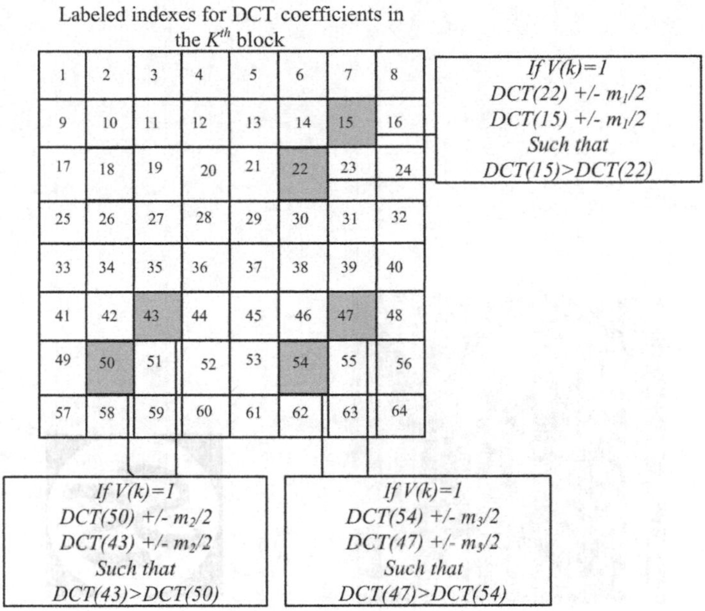

Labeled indexes for DCT coefficients in
the Kth block

1	2	3	4	5	6	7	8
9	10	11	12	13	14	15	16
17	18	19	20	21	22	23	24
25	26	27	28	29	30	31	32
33	34	35	36	37	38	39	40
41	42	43	44	45	46	47	48
49	50	51	52	53	54	55	56
57	58	59	60	61	62	63	64

If V(k)=1
DCT(22) +/- m₁/2
DCT(15) +/- m₁/2
Such that
DCT(15)>DCT(22)

If V(k)=1
DCT(50) +/- m₂/2
DCT(43) +/- m₂/2
Such that
DCT(43)>DCT(50)

If V(k)=1
DCT(54) +/- m₃/2
DCT(47) +/- m₃/2
Such that
DCT(47)>DCT(54)

FIGURE 9.4 Watermark bit embedding process for vector bit 1.

The aforementioned approach is just for motivation and is a simple approach. You can create your own efficient robust embedding approach. You can use any other transformations like DFT, DWT etc. There are even several other methods like Singular Value Decomposition (SVD) with which we can also create a robust watermarking approach. Our objective is to make an algorithm that extracts the approximate watermark bits even after a major intentional or unintentional tampering event.

9.4   EXTRACTION OF ROBUST WATERMARKS

This phase is executed at the receiver end. This phase is very important as it deals with ownership assertion. In previous chapters, we saw that an image may be tampered with either at the time of transmission or at the time of storing that image. Tampering may be either intentional or unintentional. We can only say that our algorithm is correct, if we can extract the exact watermark as inserted in the case where no alteration was made on the watermarked image. We can see in Figures 9.5 that (b) and (d) are similar watermark images if there is no tampering in (a).

FIGURE 9.5 Validation of a robust watermark technique: (a) Watermarked image, (b) Binary watermark, (c) Untampered watermarked image, (d) Extracted watermark.

Points worth remembering:
A robust watermarking approach is only valid if we extract the exact watermark original and if the watermarked image is untampered.

9.4.1 Effect of Unintentional Tampering on a Watermark

As we know that in a case of unintentional tampering, an image may be tampered with during the transmission phase. Most of the time, unintentional
tampering deals with the insertion of noise due to some small signal fluctuations during transmission. Figure 9.6 shows an example where watermarked
image (a) embedded with watermark (b) is tampered with unintentionally by
salt and pepper noise shown in (c). As per the property of good robust watermarking schemes, we must be able to extract the watermark image in as
healthy a way as possible. We can see in Figure 9.6 (d) that in spite of noise in
the watermarked image, we are still able to extract a good quality watermark
image.

9.4.2 Effect of Intentional Tampering on a Watermark

In this category, a watermarked image may be tampered with intentionally
either at the time of its storage or at the time of transmission by using a man
in the middle attack. In this case, amount of tampering may be high or low. In
Figure 9.7, an example of an intentional attack is shown in Figure 9.7 (c) where
a huge portion of the image is smoothed. In this type of attack, extraction
of a good quality watermark is a challenging task. As per the robustness of
our proposed approach, we are even able to extract good quality watermark
as shown in Figure 9.7 (d).

9.4.3 Extraction of Multiple Watermarks

We discussed in previous sections that we can embed multiple different or
identical watermarks in different places of the cover image. We do so because
if one or more than one portion of the watermarked image is entirely tampered
with then we still would be able to extract the other untampered watermarks
from the different locations of the watermarked image. We can see this scenario in Figure 9.8, here, (a) is a watermarked image and (b) represents the
four identical watermarks which are embedded at different locations of the
watermarked image. In Figure 9.8 (c), the watermarked image is tampered
with using a very high amount of smoothing which literally destroyed the
lower portion of the watermarked image. But in Figure 9.8 (d), we can see
that we still extract two healthy watermarks which are unaltered.

FIGURE 9.6 Unintentional tampering on watermarked image (a) Watermarked image, (b) Binary watermark, (c) Tampered watermarked image with salt and pepper noise, (d) Extracted watermark.

FIGURE 9.7 Intentional tampering on watermarked image (a) Watermarked image, (b) Binary watermark, (c) Tampered watermarked image with smoothing, (d) Extracted watermark.

Points worth remembering:
In the case of an ideal robust watermarking approach, the extracted watermark of the tampered image will be identical to the original watermark image.

9.4.4 Extraction-Based Categorization of Robust Watermarking

As we already know that on the basis of extraction mechanism, robust watermarks are categorized as non-blind, semi-blind and blind watermark. ownership assertion for each category is a little bit different. Here we have to make clarify that at the receiver end, we must have the original copy of watermark in order to make a comparison with the extracted one. But this does not mean that a robust watermark will always be non-blind. A watermarking approach will always be blind until or unless we do not use a watermark or cover image as an input for an extraction computation.

9.4.4.1 Ownership assertion for non-blind robust watermarks

As we know that in the case of a non-blind robust watermark, we need an untampered version of the cover or watermark image at the receiver end. Either a single or both images will be used for the extraction process of the watermark. This is why this category is least desired.

9.4.4.2 Ownership assertion for semi-blind robust watermark

In this category of robust watermark, we do not require a cover or watermark image at the receiver end. But here we need some side or abstract information of the watermarked/watermark image which contains the locations for watermark embedding. Once we identify the locations in which, watermark is embedded, then we simply extract the watermark from the frequency coefficients.

FIGURE 9.8 Example of multiple watermark extraction: (a) Watermarked image embedded with multiple identical watermark, (b) Multiple identical binary watermarks, (c) Tampered watermarked image with smoothing, (d) Extracted multiple watermarks.

9.4.4.3 Ownership assertion for blind robust watermark

In this class, neither the watermark nor the cover image is used to extract the watermark. We may only require a secret key which might be used in the embedding process. This class is the most desired approach.

9.4.5 Decryption of the Encrypted Watermark Bit Vector

Once we extract the encrypted watermark bit vector, we are required to decrypt or reshuffle it. Since encryption is done using a symmetric secret key, that key will be used for decryption of the bit vector. After getting the decrypted version of the watermark bit vector, we simply extract the first two sets of bits (as per the dimension of the watermark). Decimal values of these two bit sets are nothing but the dimension of the watermark image. After getting the dimension, just rearrange all the binary bits of the vector in matrix form. Depending upon whether the watermark image is gray or color, we need to make a set of eight or twenty-four bits because gray intensity watermarks can be represented by eight and color can be represented by twenty-four bits.

9.4.6 Watermark Comparison Parameters

Comparison of extracted watermarks with original watermarks is a very challenging task, because based only on visual comparison, we cannot judge the ownership assertion. In earlier sections we just saw an extracted watermark and decided its authenticity based only on the human visual system. If both extracted and original watermarks looked similar to HVS, then we judged the authorization. This technique is not good because it has too much human interaction. If we wish to automate the comparison process, then we need to use some mathematical parameters on the basis of which we can judge the correct authorization. We can categorize these parameters in two classes one is objective evaluation parameters (like precision, recall etc)and the other one is subjective evaluation parameters (like PSNR, MSE etc).

Points worth remembering:
The comparison process for extracted and original watermarks can be automated by objective and subjective evaluation parameters.

9.4.6.1 Objective evaluation parameters

We have already seen these parameters in greater detail in earlier chapters. We know that these parameters are well suited for comparison of two binary images. When our watermark image is binary in nature, then objective evaluation parameters will be used in order to compare extracted and original watermark images. An extracted watermark will be more authentic when its calculated objective evaluation parameters are near to its ideal values.

Points worth remembering:
Objective evaluation parameters are well suited for binary watermarks.

9.4.6.2   Subjective evaluation parameters

Subjective evaluation parameters are well suited for comparison of two gray scale or two color images. When our watermark image is gray scale or color, then objective evaluation parameters may give misleading results, hence subjective evaluation parameters are used in order to compare extracted and original non-binary watermark images. The extracted watermark will be more authentic and healthy when its calculated subjective evaluation parameters are near to its ideal values.

Points worth remembering:
Subjective evaluation parameters are well suited for gray scale and color watermarks.

SUMMARY

– *Robust watermarks* are used for ownership assertion and to prove authorization.

– Robust watermarks are *more resistant against various types of intentional and unintentional attacks.*

– The *self embedding approach* is highly unsuitable for robust watermarking which is why we use an image as watermark that is related to the owner of the cover image .

– On the basis of extraction, robust watermarking is classified in three ways: *non-blind, semi-blind and blind robust watermarking.*

– *Binary images* are most the suitable type of watermarks for use in robust watermarking as they take very few bits to represent the principal content of the watermark.

– An *encrypted watermark* can be embedded in order to avoid malicious embedding of an authentic watermark into the objectionable image.

– For the embedding of robust watermarks, the *frequency domain* is more suitable than the spatial domain.

– There are various frequency domain transformations available like DFT, DCT, DWT etc which can be utilized for watermark embedding in a robust watermarking approach.

– *A robust watermarking approach is only valid* if we extract the exact watermark as the original if the watermarked image is untampered.

– In the case of an *ideal robust watermarking approach,* the extracted watermark of the tampered image will be identical to the original watermark image.

– The comparison process for extracted and original watermarks can be automated by *objective and subjective evaluation parameters.*

– *Objective evaluation parameters* are well suited for *binary watermarks.*

– *Subjective evaluation parameters* are well suited for *gray scale and color watermarks.*

Dual Watermarking

CONTENTS

So far, we have seen fragile watermarking with only alteration detection capabilities, fragile watermarking with recovery capabilities in the spatial as well as frequency domains and the robust watermarking approach. All the approaches discussed have different applications and principles. Sometimes we require an approach which has functionality in both fragile and robust types of approaches. This type of watermarking approach can be referred to as a dual watermarking approach. Suppose we have an image on our website and before uploading it, we want to protect it in such a way that at the time of a conflict, we can show the authentication (if tampered) for that image as well as the authorization. Authentication of the image can be tested if it is protected with a fragile watermark and authorization can me judged if it is protected with robust watermark. This scenario motivates us to develop a new watermarking approach which is the combination of both approaches. We can also enhance the functionality of the dual watermarking approach by making it recoverable i.e. if any tampering is done, we must be able to recover the principal content of the image.

Hence in this chapter, we deal with various theoretical and practical aspects of dual watermarking schemes dealing with authentication or integrity as well as authorization of the associated cover images. The **Chapter Learning Outcomes(CLO)** of this chapter are given below. After reading this chapter, readers will be able to:

1. Understand the background and need for dual watermark.

2. Understand the different terminologies related to dual watermarks.

3. Write their own dual watermarking algorithms.

4. Identify the research gap and research interest in this field.

10.1   INTRODUCTION

A dual watermark is a mark which has functionality for both types, namely, fragile as well as robust watermarks. As we know that a fragile watermark has the capability detecting the smallest change in the image. If a given fragile watermark is more advanced, then it can also recover the tampered content of the image. On the other hand, robust watermarks are entirely different from fragile watermarks. They can tolerate intentional and unintentional attacks so they used for copyright protection of a watermarked image. Since both types of watermarks are very opposite in nature, developing an algorithm which fulfils the requirements of both is a very difficult task. In earlier chapters, we studied how the frequency domain is well suited for robust watermark whereas spatial domain is suitable for fragile watermarking. In this chapter, we inherit all the concepts of robust and fragile watermarks in order to create dual watermarks. We just need to pay attention to the embedding sequence of both watermarks. Though this technique is complex, it is really useful for multiple purposes.

Points worth remembering:
Dual watermarks have functionality of both types of watermarks viz robust and fragile.

10.1.1   Need for Dual Watermark

Since dual watermarks have the functionality of both fragile and robust watermark, they are applicable in all those areas where these watermarks can be used independently. At the same time, they can also be utilized in all areas where both are required simultaneously. We can understand them from an example. Suppose we are maintaining a website which deals with highly sensitive and confidential images. These images can only be owned by the website owner. In the future, due to lack of security, it may be possible that a intruder could steal confidential images and alter them according to his wishes and upload them somewhere else. When this crime is brought to our attention, at that time we have two challenges. One is to prove authorization of the images and the second is to prove that the image has been altered or not. A richer dual watermark will also be able to recover all the tampered contents.

Points worth remembering:
An ideal dual watermark can provide ownership assertion, content authentication and recovery of tampered principal content of the image.

10.1.2   Managing Robustness and Fragility at the Same Time

Till now it was understood that robustness and fragility are very opposite properties. In the case of robustness, the watermark must survive in case of an intentional or unintentional attack whereas in the case of fragility, the watermark must be destroyed even through a change in a single pixel intensity.

When we say that a dual watermark has robust and fragile properties, it looks somewhat impractical, but this is true and is the beauty of dual watermarks. It is made in such a way that it can act as a robust watermark as well as a fragile watermark simultaneously. Whether a watermark is robust or fragile can only be judged by its generation and embedding processes. Both watermarks are generated and embedded by their own specified techniques. We can say that there is no dependent relationship between the performance of the robustness and fragility for dual watermarks. That is why dual watermarks hold properties of both. We will see the entire generation and embedding process of dual watermark in more detail in the next section.

Points worth remembering:
There is no trade-off between robustness and fragility in dual watermarks. This means that both properties can be increased or decreased simultaneously.

10.1.3   Image for Robust Watermark and Self-Embedding for Fragile Watermark

In earlier chapters, we saw that a watermark could be either a different image which could be meaningful in nature or could be a matrix of random bits generated from the cover image itself which is meaningless in nature. Generation of random bits using the various properties of the cover image is called a self-embedding approach. In self-embedding, generated watermark bits are highly associated with the cover which is why the smallest change can be noticed. Thus self-embedding is the best method for fragile watermarking. On the other hand, robustness can be achieved by effectively embedding the separate watermark in the cover image. Robustness is not a property of the watermark but of the embedding process. That is why we use an image which is related to the owner of the cover image like the copyright logo as a watermark. A watermark which is embedded using robust embedding is called robust watermark. In case of dual watermarking, we do the same job. A separate image will be used for the robust watermark whereas self-embedding will be used for a fragile watermark in dual watermarking. We have to pay attention to the sequence of the embedding of robust and fragile watermarks. We will see the need for a proper sequence in embedding in the next section in detail.

Points worth remembering:
In dual watermarking, a separate image is used for a robust watermark while self-embedding is used for a fragile watermark.

10.1.4 Achievable Security Requirements through Dual Watermarking

There are some security requirements which must be fulfilled for any efficient cryptographic approach, namely, authentication, integrity, authorization, confidentiality and non-repudiation. Since watermarking is a kind of security approach, we need to cover as much as security requirements as possible. Lets discuss all security requirements briefly. Authentication is the property in which we make sure that the receiver of the message is the person for which that message is intended. Integrity of a message refers to protecting the message from being modified by unauthorized parties. When we talk about the confidentiality of a message, we are talking about protection of the information from disclosure to unauthorized parties. Authorization is nothing but access control over the message. Non-repudiation is a service that provides proof of the integrity and origin of the data. A default dual watermarking approach achieves maximum security requirements like authentication, integrity, confidentiality and authorization which is not possible in the case of individual robust or fragile watermarking schemes. Our objective is to make an algorithm that provides maximum security requirements. Besides all of these properties, dual watermarking may also provide recoverability of tampered content.

Points worth remembering:
Dual watermarking can achieve maximum security requirements compared to individual robust or fragile watermarking schemes.

10.1.5 Semi-Fragile Watermarks vs. Dual Watermarks

In an earlier chapter, we discussed semi-fragile watermarking schemes. One should not be confused between semi-fragile watermarking and dual watermarking schemes. Each is very different. A semi-fragile watermark also has properties of both robust and fragile watermarks but at a time it acts as either robust or fragile. There is a threshold value for tampering, and if tampering of the image is below that threshold value, then it acts like a robust watermark. When tampering exceeds that threshold value, then watermark acts fragile. We can simply say that for routine processing of an image, whatever copyright information is embedded will be safe. If an intentional attack is done on the image, then the copyright image will be destroyed and it will show the tampered regions. Dual watermarking is totally different because here we can enjoy the properties of both robust as well as fragile watermarks simultaneously.

Points worth remembering:
A semi-fragile watermark and a dual watermark each are very different.

10.1.6 Types of Dual Watermarks

Categorization of dual watermarks can easily be understood from Figure 10.1. Here we see that dual watermarks can be broadly classified into two types:

1. On the basis of the embedding mechanism.

2. On the basis of the extraction mechanism.

10.1.6.1 On the basis of the embedding mechanism

When a dual watermark is generated, it must be embedded into the cover image in such way that it can effectively do its assigned job (authentication, authorization and tamper recovery). On the basis of job assigned to the watermark, it is further classified into two categories:

1. **Dual watermark with authentication and authorization:** In this category, a dual watermark will provide ownership assertion with tamper detection. Since here we are not performing content recovery, this approach is less complex. Embedding as a robust watermark will always be on a block basis, hence we can further categorize this approach on the basis of authentication. Authentication or we could say tamper detection may be either pixel-based or block-based.

   Block-based tamper detection: As its name suggests, in this category, a watermark is generated for the blocks of the given cover image. First of all the cover image is divided into the non overlapping blocks of fixed size. After that, the generated watermark is embedded into each block. The main advantage of this type of dual watermarking scheme is its low complexity because we do not need to process each and every pixel. We treat the whole block as a single pixel. Block-based tamper detection is less accurate because if a single pixel of the given block is altered, then the whole block will be treated as an altered block, in spite of having some pixels of that block unaltered. Hence this type of watermark is unsuitable for exact localization of tampering. This scheme is useful when we only need to know whether the given watermarked image in authentic or not. Here our assumption is that a robust watermark has been

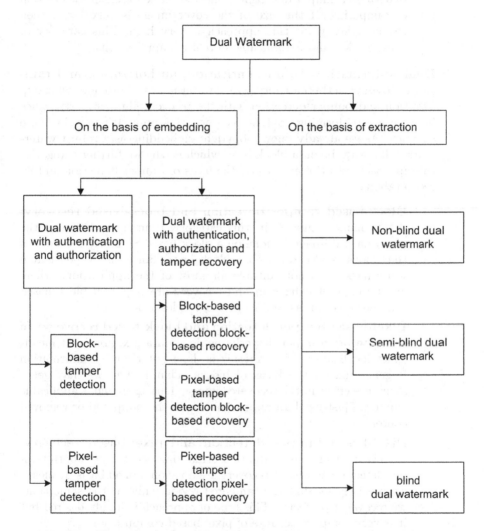

FIGURE 10.1 Classification of dual watermark.

embedded on a block basis. That is why we are not taking it into consideration for categorization.

Pixel-based tamper detection: In this category, a watermark is generated using each pixel of the given cover image. That is why pixel-based tamper detection is most suitable for exact localization of tampering. If the size of the cover image is very large, then the complexity for this approach is very high. This category of watermark is more applicable in real life applications.

2. **Dual watermark with authentication, authorization and tamper recovery:** In this category, a dual watermark will provide ownership assertion and tamper detection with the recoverability property. Since here we are performing content recovery, this approach may be more complex. As we already know, copyright embedding as a robust watermark will always be on a block basis which is why we further categorize this approach into three types on the basis of tamper detection and its recoverability.

Block-based tamper detection and block-based recovery: As its name suggests, in this category, tamper detection of the watermarked image is done on a block basis. Similarly recovery of that block is also done on a block basis. Since this type of dual watermarking is not suitable for most of the applications where exact tamper localization and recovery are required, block-based tamper detection as well as recovery is least desired.

Pixel-based tamper detection and block-based recovery: In this category, tamper detection is done on a pixel basis. It means exact localization of tampering is done but recovery information is generated on the basis of blocks. That is why we may get a stair case effect in the recovered image. This approach is applicable where we just need an approximation of the tampered or removed content.

Pixel-based tamper detection and pixel-based recovery: This is an ideal case for dual watermarking because here both tamper detection as well as recovery are done on a pixel basis. It means we precisely localize the tampered region while at the same time, we recover it perfectly. This type of approach is highly desired but it is very complex because of pixel-based computation.

Points worth remembering:
For dual watermarking, our assumption is that robust watermarking will always be done on a block basis.

10.1.6.2 On the basis of the extraction mechanism

At the receiver end, once the watermarked image is received, we need to extract the watermark in order to check the integrity of the watermarked image for any further application. On the basis of watermark extraction, a dual watermark is further classified into three categories:

1. **Non-blind dual watermark:** In this category of dual watermark, we require the original cover image for tamper localization and original cover or watermark for ownership assertion at the receiver end. This class of watermark is not at all desirable because it increases the overhead for storing and transmitting the cover image/watermark along with the watermarked image.

2. **Semi-blind dual watermark:** In this category of dual watermark, instead of the whole cover image and/or watermark, we require only some side and abstract information at the receiver end, to detect the tampered region as well as for ownership assertion. Sometimes we embed the watermark in some region-of-interest (ROI) of the cover image (See in detail in the next section). In such a case, we need to keep track of the embedding locations of the watermark in the cover image. This track record can be treated as side or abstract information of the cover image and it is required at the receiver end in order to detect the tampering of pixels. We need to transmit this side information over a secure channel or this information must be encrypted to prevent it from any middle attack.

3. **Blind dual watermark:** This category of dual watermark is most desirable because it does not require any information related to the cover image or watermark at the receiver end during tamper detection or the ownership assertion process. Watermark embedding and extraction algorithms are well known to all, so to prevent the unauthorized extraction or embedding of invalid watermarks into a cover image, we need a some symmetric key known by only the sender and receiver. So, in the case of a blind dual watermark, we need only the symmetric key at the receiver end that was used during embedding of the watermark.

Points worth remembering:
There should be a secure mechanism to transmit the auxiliary information required for extraction of the watermark at the receiver end.

10.2 GENERATION AND EMBEDDING OF DUAL WATERMARKS

In earlier chapters, we studied generation and embedding of the watermark in two different sections but here we are going deal with generation and embedding in a single section. The reason behind this is that in the case of a fragile watermark, the generation process plays an important role whereas in the case of robust watermarking, the embedding process plays an important role. Since in dual watermarking, both watermarking approaches are included, we need to discuss both embedding and generation mechanisms in parallel. When we talk about fragile watermarking approaches, then we generate watermarks with the help of a self-embedding process and the embedding process is very straightforward in the spatial domain. We can say that in fragile watermarking, generation of the watermark is a primary process whereas embedding is a secondary process. When we talk about a robust watermarking approach, then we do not need to generate a watermark because we use a separate image as the watermark, but we need to embed it in an efficient manner so that it remains robust against various attacks. We can say that in the case of robust watermarking, generation of watermarks is secondary whereas embedding is a primary process.

10.2.1 Hierarchical Approach for Watermark Generation and Embedding

In the previous section, we saw that both watermarks (robust and fragile) which constitutes a dual watermark, are different in nature as well as different with respect to the generation and embedding point of view. Here the main questions that arise are how to embed/generate the watermarks? which watermark will be embedded first? What will be the effect of one watermark on another? How should the imperceptibility between the cover and watermarked image as well as between the original watermark and extracted watermark be maintained?

To answer all these questions we need to perform the generation/embedding process of the watermark in a hierarchical manner as shown in Figure 10.2. This hierarchy must be in such a way that it minimizes the overhead for embedding/generation and also minimizes the false acceptance rate and false detection rate. We will see all these terms one by one in later sections. First, we need to find a proper sequence for generation and embedding of both watermarks. A efficient sequence may be as follows:

1. Embedding of the watermark (selected as copyright) in the frequency domain.

2. Fragile watermark generation using a self-embedding approach from the copyright embedded image.

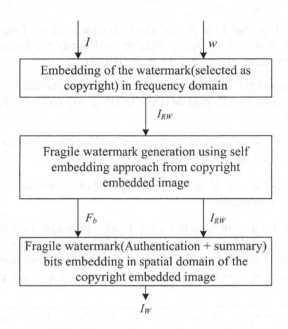

FIGURE 10.2 Hierarchical approach for watermark generation and embedding.

3. Fragile watermark (Authentication + summary) bit embedding in the spatial domain of the copyright embedded image.

Points worth remembering:
A dual watermarking approach always follows a fixed sequence of embedding (robust then fragile).

The given hierarchical sequence shows three steps for dual watermark generation and embedding. Now we will see the significance of each step in detail.

10.2.1.1 Embedding of the watermark (selected as copyright) in the frequency domain

It is clear that we will embed the robust watermark first. To execute this step we need two inputs; first one is the cover image and the other one is the watermark because in the case of robust watermarking, we use the watermark explicitly. Let us consider one of the simplest methods in order to understand the whole process of generation and embedding of dual watermarks. Let the cover image be I of size $M \times N$ and the watermark be w of size $m \times n$. Here our assumption is $m \times n = \frac{M \times N}{64}$. Now we will apply the following steps for embedding the watermark w in I to make it robust.

1. As we know that default robust watermarking is block-based which is why first of all we divide the cover image into non-overlapping blocks of size 8×8.

2. Convert the watermark w into a vector V of size $1 \times \frac{M \times N}{64}$ using either a row wise or column wise scan.

3. Encrypt the vector V by using a symmetric secret key S_k in order to provide confidentiality.

4. Now each bit of encrypted vector V will be embedded into the corresponding 8×8 block of the image I. As we know that in the case of robust watermarking, embedding is done in the frequency domain because frequency coefficients are more robust than the pixel intensity itself.

5. The aforementioned steps are preliminary steps which must be followed. After that we need to create an efficient algorithm using any transformation such as DCT, DFT, DWT etc.

6. Suppose for this example we use a DCT transformation for embedding the encrypted watermark bits in the blocks of image I

$$I_{RW} = Robust\ Watermarking(DCT, V, I) \quad\quad (10.1)$$

Here I_{RW} is a partially watermarked image embedded with a robust watermark. *Robust Watermarking()* is a self generated function which takes a three argument transformation function, encrypted watermark bits V and cover image I as inputs. Now the partially embedded image I_{RW} will be treated as input for the next step in fragile watermark generation.

Points worth remembering:
First of all an image explicitly taken as the watermark is embedded into the frequency domain of the cover image.

10.2.1.2 Fragile watermark generation using a self-embedding approach from the copyright embedded image

In this step, we treat partially watermark embedded image I_{RW} as the cover image and generate the fragile watermark bits which include authentication as well as summary bits. These bits may be generated on a pixel or block basis as follows:

1. If we want to generate a fragile watermark for block-based authentication and block-based recovery, then first of all divide the image I_{RW} into non-overlapping blocks of size $i \times j$. Here i and j are decided as per the level of complexity and accuracy of the result. If i and j are small, then the algorithm will be complex, but the result of tamper detection and recovery will be fine and vice versa.

2. Now we need to generate fragile watermark bits for each block independently. Fragile watermark bits consist of authentication bits and summary bits.

3. In earlier chapters we already saw various self-embedding methods to generate authentication bits either based on pixel position or its relation with neighbors etc. We can choose any one of them or we can make a new one. Suppose the vector of the generated authentication bits for image I_{RW} is denoted by A_b.

4. If we only deal with authentication and authorization of the image without any tamper recovery, then A_b will be treated as fragile watermark bits.

5. In order to increase the security, we can encrypt A_b using a symmetric key S_k.

6. If the vector for fragile watermark bits, are denoted by F_b, then

$$F_b = Encryption(A_b, S_k) \tag{10.2}$$

7. Most of the time, besides authentication and authorization properties, we also require tamper recovery. In earlier chapters we saw various self-embedding methods to generate summary bits such as those based on mean values of the block or any cluster-based method which is required for tamper recovery. We can choose any one of these methods or we can develop new ones as per the requirement. Suppose the vector of the generated summary bits for image I_{RW} is denoted by S_b.

8. Now the fragile watermark bit vector F_b consists of both of the bit vectors A_b and S_b, hence we can denote it as

$$F_b = Encryption(A_b, S_b, S_k) \tag{10.3}$$

9. Similarly if we want to generate a fragile watermark for pixel-based authentication and block-based recovery or pixel-based authentication and pixel-based recovery, then we need to process the image I_{RW} accordingly in order to generate F_b.

10. Our next objective is to embed the fragile watermark bits F_b into cover image I_{RW}.

Points worth remembering:
A copyright embedded image is treated as input for the next step in fragile watermark generation.

Points worth remembering:
A fragile watermark (Authentication and/or summary bits) is generated using self-embedding approach.

Points worth remembering:
Dual watermarks may be with or without a tamper recoverability capability.

10.2.1.3 Fragile watermark vector F_b embedding in the spatial domain of the copyright embedded image I_{RW}

Once we generate the fragile watermark bit vector F_b then we need to embed it in a way to retain high fragility. In earlier chapters, we saw that in order to provide fragility, we need to embed the watermark in the spatial domain so that the smallest change in the image could be detected. We may therefore proceed in the following steps:

1. Fragile watermark vector F_b contains block wise authentication bits A_b and summary bits S_b for the entire image I_{RW}. First of all we divide the fragile watermark vector F_b into partitions where each partition contains fragile watermark bits for the corresponding block.

$$F_b = F_b \text{ bits for } 1^{st} \text{ block} + F_b \text{ bits for } 2^{nd} \text{ block} ..... + F_b \text{ bits for} (\frac{M \times N}{64})^{th} \text{ block}$$

2. Map the watermark bits of one block to another block using a secret key. Let us consider that the fragile watermark bits of B_i^{th} block are mapped with LSBs of the pixel intensities of the B_j^{th} block and vice versa. So, if block B_i^{th} is tampered with then its recovery information will be preserved in B_j^{th} block.

3. Since these fragile watermark bits will be embedded in the spatial domain of I_{RW}, hence we need to decide how many bits should be dedicated for

embedding. This decision is also important because non uniform assignment of watermark bits may degrade the image quality as well as decrease the imperceptibility.

4. This fragile watermark embedded image is the final watermark image denoted as I_w

$$I_w = Watermark\ Embedding(F_b, I_{RW}) \qquad (10.4)$$

A graphical example for the above mentioned valid hierarchical approach of a watermark embedding process is shown in Figure 10.3 where Figure 10.3(a) is the cover image I, (b) is the copyright logo which will serve as binary watermark w, (c) is the partially watermarked image I_{RW} embedded with w, (d) is the fragile watermark vector F_b generated using self-embedding approach, and (e) is final watermarked image I_w which is obtained after embedding of F_b in I_{RW}. We can see here that there is no noticeable distortion in images (a), (c) and (e). This means the imperceptibility between partially/ completely embedded watermarked images and the cover image is very high which is one of our objectives.

Points worth remembering:
A fragile watermark is embedded into the spatial domain of the robust watermark embedded image.

Points worth remembering:
A fragile watermark is created on a block mapping basis.

10.2.2   Validity of the Proposed Dual Watermark Embedding Sequence

In the above section, we saw a hierarchical approach for dual watermark embedding. Now a question arises about how we can say that the embedding of a robust watermark followed by a fragile watermark is valid. Or what will happen if we change the embedding sequence (fragile then robust)? Let us consider a scenario where we change the embedding sequence.

Suppose first of all, fragile watermark vector F_b is generated from the cover image I using a self-embedding approach. Now this generated fragile watermark will be encrypted using a secret key S_k and then embedded into the spatial domain (LSBs of the pixel intensities) of the cover image I.

$$I_{FW} = Watermark\ Embedding(F_b, I, S_k) \qquad (10.5)$$

Here I_{FW} is the fragile watermarked embedded image and $Watermark$

FIGURE 10.3 Example of valid hierarchical approach for watermark embedding process: (a) Cover image I, (b) Copyright logo as binary watermark w, (c) Partially watermarked image I_{RW} embedded with w, (d) Fragile watermark bits F_b generated using self-embedding approach, (e) Final watermarked image I_w obtained after embedding F_b in I_{RW}.

$Embedding(F_b, I, S_k)$ is the self-embedding function which takes the three argument fragile watermark F_b, cover image I and secret key S_k. This partially watermarked image will be passed to the next step as an input image. In the next step, I_{FW} will be treated as cover image and any explicitly chosen watermark w as the copyright will be embedded into the frequency domain of I_{FW}. The robust watermarking approach will look like

$$I_w = Watermark\ Embedding(Transformation, I_{FW}, w) \qquad (10.6)$$

Here I_w is the final watermarked image and $Transformation$ may be any frequency domain transformation like DCT, DFT, DWT etc which is applied on I_{FW} in order to embed w.

Now the interesting point is here is that when we embed any copyright information in the frequency domain of the image as a robust watermark, then it definitely will change the actual intensity of the cover image. This change is very small which is why the human visual system cannot notice it but our embedded fragile watermark can easily detect these changes as it is intended to do so. Now the problem is when at the receiver end we extract the fragile watermark first, then it will show most of the intensities as tampered only because of embedding of the robust watermark. Actually this tampering result is false because changes due to embedding of copyright information are obvious and can be ignored. If there are actual changes in the watermarked image because of an attack, then we cannot distinguish between actual and false changes. So we can say if we change the sequence of embedding for dual watermarks then we can be the victim of high FAR and FRR.

Points worth remembering:
Embedding of a fragile watermark first then a robust watermark may cause high FAR and FRR.

One more doubt may arise here that when we embed fragile watermark F_b in copyright embedded image I_{RW} (valid sequence) then the intensities of the pixels of I_{RW} may also change because of changes in the LSBs. In this case, when we extract the copyright image at the receiver end, will the extracted copyright be distorted or not? Actually robust watermarking is made in such a way that it resists various intentional and unintentional attacks. A change in LSBs due to embedding of a watermark is also a kind of minor attack. If our robust watermark embedding approach is good enough to tolerate this kind of alteration, then the extracted copyright at the receiver end will not be distorted and looks healthy.

The aforementioned reasoning can be understood from Figure 10.4. Here an example has been taken of watermark embedding and extraction processes when no intentional or unintentional attack is done on watermarked image I_w and valid hierarchy of embedding is used. In Figure 10.4, (a) is a cover image I which is going to be embedded with copyright logo (b) serving as binary watermark w. Figure 10.4 (c) is partially watermarked image I_{RW} embedded with w, and (d) is fragile watermark vector F_b generated using any self-embedding approach. (e) is final watermarked image I_w obtained after embedding of F_b in I_{RW}. These images from (a) to (e) are nothing but the embedding results of any dual watermarking approach. Suppose that we are not tampering with the watermarked image I_w at all, now we have to see what will be the extraction result. Figure (f) is the extracted fragile watermark for tamper detection or authentication. Here the black portion shows the untampered region. We can see that (f) is a completely black image, and it means that FAR is zero in this case. Now we see what will happen with the extracted copyright image. Figure (g) is the extracted copyright watermark image w. We can observe that (g) contains some noise with negligible density. This is only because of the change in the image I_{RW} after embedding F_b. The interesting thing is here is that we have some scope to extract completely noiseless watermark image w because its all about the robustness of the approach used to embed w in I.

Figure 10.5 shows an example of watermark embedding when a valid sequence is not used. In Figure 10.5, (a) is the over image I and (b) is the fragile watermark bits F_b generated using any self-embedding approach. Now instead of embedding watermark w first, we will embed the fragile watermark bits first and the partially watermarked image I_{FW} embedded with F_b is shown in (c). Figure (d) is the binary watermark w used as copyright image. The final watermarked image I_w which is obtained after embedding of w in I_{FW} is shown in the figure (e). From this example, we can conclude after seeing images (a),(c) and (e) that there are no significant visual changes observed in the watermarked images if we change the embedding sequence of w and F_b. A change in the embedding sequence causes an incorrect result during the extraction process. We can see it in Figure 10.6, where we see that first of all we are embedding the fragile watermark bit F_b shown in (b) into the cover image and the resultant partial watermark image I_{FW} is shown in (c). Figure (d) is the binary watermark w used as copyright image. The final watermarked image I_w which is obtained after embedding of w in I_{FW} is shown in the figure (e). As no change (intentional or unintentional) is done on the image I_w after the embedding of w, the extracted watermark w will be completely noiseless as desired. A problem will arise with the authentication of the image because embedding of w actually changes the intensities of I_{FW} unknowingly and as per the property of fragile watermarks, these changes will be treated as an attack. That is why the figure (g) is not completely black, and white spots show the locations where pixel intensities are changed due to embedding of w. This result may mislead the receiver because of non-zero FAR. For this reason we can say that this embedding sequence must be avoided.

10.3 EXTRACTION OF DUAL WATERMARKS

The extraction phase is executed at the receiver end. This phase consists of three steps as follows:

1. Extraction of copyright information which is embedded as a robust watermark, in order to ensure authorization.

2. Extraction of fragile watermark to verify that the received image is authentic or not?

3. If the image is not authentic and tampered with then this step is executed for recovering the tampered regions.

In the aforementioned steps, third step will always come after the second step for reducing the complexity. If the image is not tampered with, then there is no need to extract the recovery bits and process it. One more order of execution may also

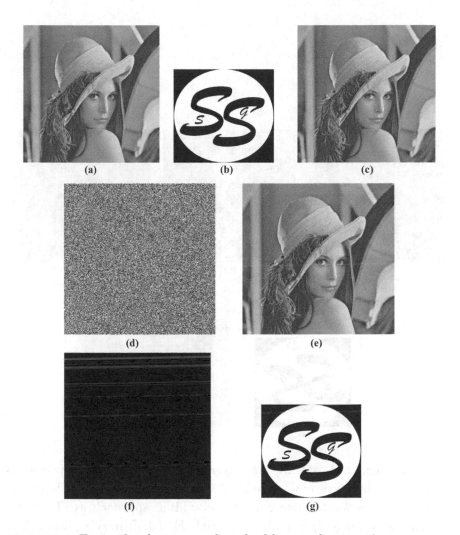

FIGURE 10.4 Example of watermark embedding and extraction processes when no attack is carried out and an valid hierarchy is used: (a) Cover image I, (b) Copyright logo serving as binary watermark w, (c) Partially watermarked image I_{RW} embedded with w, (d) Fragile watermark bits F_b generated using self-embedding approach, (e) Final watermarked image I_w embedded, obtained after embedding of F_b in I_{RW}, (f) Tamper detection result using a fragile watermark, (g) Extracted watermark w.

FIGURE 10.5 Example of watermark generation and embedding processes when an invalid hierarchy is used: (a) Cover image I, (b) Fragile watermark bits F_b generated using a self-embedding approach, (c) Partially watermarked image I_{FW} embedded with F_b, (d) Binary watermark w used as the copyright image, (e) Final watermarked image I_w obtained after embedding of w in I_{FW}.

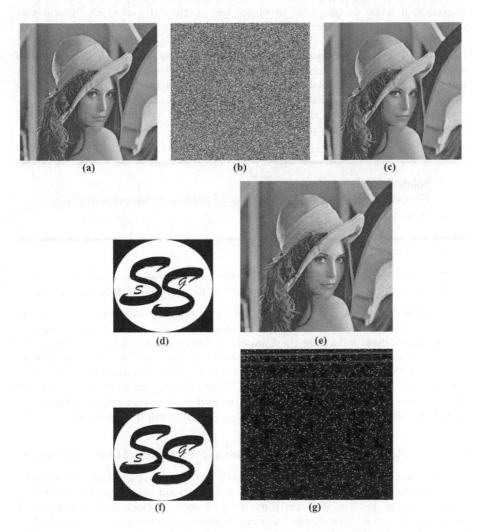

FIGURE 10.6 Example of watermark embedding and extraction processes when no attack is carried out and an invalid hierarchy is used: (a) Cover image I, (b) Fragile watermark bits F_b generated using self-embedding approach, (c) Partially watermarked image I_{FW} embedded with F_b, (d) Binary watermark w used as the copyright image, (e) Final watermarked image I_w obtained after embedding of w in I_{FW}, (f) Extracted watermark w, (g) Tamper detection result using a fragile watermark.

be valid such as second then third and finally, the first step. In the case of an ideal approach, if image is tampered in major region, then the copyright information will also be intact as embedded. In the previous chapter, we studied where in case of robust watermarking, if there is no alteration done on the watermarked image, then we will extract the same copyright information as embedded. In the case of dual watermarking, this scenario may not be true, if a robust algorithm is less robust. Changes done by embedding of a fragile watermark must be ignored in any developed algorithm. We saw in the first section that on the basis of the extraction mechanism, a dual watermark is categorized into three types and authentication and authorization processes are different in these.

Points worth remembering:
The tamper detection phase will always be followed by tamper recovery in order to reduce complexity.

10.3.0.1 Authentication and authorization for non-blind dual watermark

As we know that in the case of non-blind dual watermarks, we need an untampered version of watermarked and copyright images also at the receiver end. Pixel wise comparison is done to localize the tampering. Sometimes there is some tampering which cannot be identified just by our human visual system, and for that case we require pixel wise comparison of untampered and tampered watermarked images in this category. Actually the original version of the copyright image is always required at the receiver end for comparing it with the extracted watermark, but if it is also used in the extraction process of the copyright, then it will be non-blind.

10.3.0.2 Authentication and authorization for semi-blind dual watermarks

In this category of tamper localization, we do not require original watermarked or copyright images at the receiver end. But here we need some side or abstract information of watermarked and watermark images containing the locations for watermark embedding. Once we identify the locations in which a watermark is embedded, then we simply extract the watermark from the LSBs. For tamper detection, we again generate the fragile watermark using same algorithm by which it was generated at the time of embedding. Now we have two sets of fragile watermarks, one is the extracted one and the other one is the recalculated one. We simply do pixel wise comparison of these two watermarks, and if any mismatch is found between the recalculated and extracted one, then that pixel is marked as tampered pixel. In the case of authorization, we extract the copyright information from the locations available in side information and compare it with the original one.

10.3.0.3   Authentication and authorization for blind dual watermark

In this class, the same method as used in semi-blind fragile watermarks, is applied for tamper localization and ownership assertion. The only difference is that here in this case of fragile watermark, we extract and recalculate the fragile watermark for the entire set of pixels of the image. In this category, we do not need any side information related to locations, because the watermark is embedded in all pixels. We just need a symmetric key at the receiver end by which watermark is embedded. The same process is also applied for ownership assertion. The copyright is extracted from the watermarked image using the symmetric key and compared with the original copyright.

Tamper detection and recovery results are affected by the way of embedding of the dual watermark. As we know that a robust watermark is always embedded block wise but fragile watermark can be embedded either block wise or pixel wise. Hence we can find the accuracy accordingly.

10.3.0.4   Block wise tamper detection

If the fragile watermark of a dual watermark is embedded using a block wise embedding technique then we will not be able to determine the exact altered pixels. Actually in this case the image is divided into non-overlapping blocks and the cumulative fragile watermark bits are generated using all pixels of that block. These generated fragile watermark bits are embedded into the parent block or any other mapped block. If any intentional or unintentional tampering is done on any pixel of a block, the entire block will be treated as altered. In Figure 10.7, the image of Lena is watermarked with a block-based self-embedding approach and some alteration is performed on the watermarked image in Figure 10.7 (c). But we can see in Figure 10.7 (d) that the whole block is treated as tampered which contains those pixels also, which are not actually tampered. Copyright information is extracted and shown in Figure 10.7 (e).

10.3.0.5   Pixel wise tamper detection

If the fragile watermark of a dual watermark is embedded through a pixel wise embedding technique, then we can do exact localization of the alteration in images. Here each pixel stores its own fragile watermark in its LSBs and if any tampering is done, then the extracted and recalculated LSBs will not match. This type of detection is called pixel wise detection. In Figure 10.8, the image of Lena is watermarked with a pixel-based self-embedding technique. An intentional attack is done on the lena image in which a rose is added on the hat of lena. At the receiver end, tamper localization process is performed. Figure 10.8 (d) shows the tamper localization result. Here white pixels show the tampered region, whereas black pixels show the untampered region. We can see that exact localization of tampered pixels is possible in a pixel-based self-embedding approach.

10.3.1   Tamper Recovery in Dual Watermarking

Till now we have seen the tamper detection results for block as well as pixel-based fragile watermarking methods in dual watermarks. Tamper recovery is also one of

FIGURE 10.7 Example of block-based embedding of fragile watermark:
(a) Original image, (b) Copyright image, (c) Tampered version of (a),
(d) Block-based tamper detection, (e) Extracted copyright image.

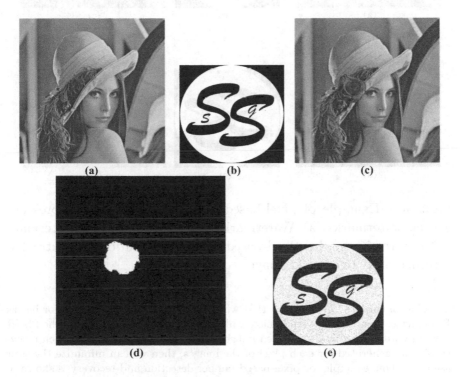

FIGURE 10.8 Example of pixel-based embedding of a fragile watermark:
(a) Original image, (b) Copyright image, (c) Tampered version of (a),
(d) pixel-based tamper detection, (e) Extracted copyright image.

FIGURE 10.9 Example of pixel-based tamper recovery in the presence of dual watermark: (a) Watermarked image, (b) Binary watermark (c)Tampered image, (d) Pixel-based tamper localization, (e) Extracted watermark, (f) Recovered image.

the points of concern. As we know that when recovery bits are generated for blocks of the image, then the recovered image will contain a stair case effect and the PSNR between watermarked and recovered images will be very low. Similarly when recovery bits are generated for each pixel of the images, then we can minimize the stair case effect. One example for pixel-based tamper detection and recovery is shown in Figure 10.9. Here we can see that the tamper detection and recovery results are very accurate.

Points worth remembering:
The quality of the tamper recovery is based on the method by which the watermark is embedded (either block or pixel wise).

10.3.2   Tamper Detection and Watermark Comparison Parameters

The efficiency of a dual watermark can be judged by the combined accuracy of the fragile watermark and robust watermark. Since both watermarks have very opposite qualities, the comparison parameters for both will be different. For fragile watermarking, tamper detection at the receiver end can be judged by two well-known parameters which are False Acceptance Rate (FAR) and False Rejection Rate (FRR) which we have already seen in the earlier chapters.

For robust watermarking, comparison of the extracted watermark with the original watermark is a very challenging task, because we cannot judge the ownership assertion only on the basis of visual comparison. In earlier chapters, we saw that besides the human visual system, we need some objective evaluation parameters (precision, recall etc) in order to compare the extracted and embedded watermarks. Generally, we use binary images as watermarks, hence objective evaluation parameters are well suited for that. If a watermarked image is tampered with, then we need to recover those tampered portions. A watermark may be gray scale or a color image, hence we need subjective parameters (PSNR, SNR etc) in order to compare the original watermarked image and its recovered version.

Points worth remembering:
In the case of robust watermarking, the comparison process of extracted and original watermarks can be automated by objective evaluation parameters.

Points worth remembering:
In the case of fragile watermarking, accuracy is measured using FAR and FRR metrics

---

Points worth remembering:
If tamper recovery is done, then imperceptibility measurements between the original watermarked and the recovered image is done using subjective evaluation parameters.

SUMMARY

- A *dual watermark* has the functionality of both types of watermarks viz robust and fragile.

- An ideal dual watermark can perform *ownership assertion, content authentication and recovery of tampered principal content* of the image..

- There is *no trade-off* between the robustness and fragility of a dual watermark. It means that both properties can be increased or decreased simultaneously..

- In dual watermarking, a separate image is used for robust watermark and self-embedding is used for fragile watermark.

- Dual watermarking can achieve the *maximum security requirements* in comparison to individual robust or fragile watermarking schemes.

- A *semi-fragile watermark* and a dual watermark both are entirely different.

- For dual watermarking, our assumption is that robust watermarking is done on a block basis.

- A dual watermarking approach always follows a *fixed sequence of embedding (robust then fragile)*.

- A Dual watermark may be with or without tamper recovery capabilities.

- Embedding of a fragile watermark before a robust watermark may cause nonzero FAR and FRR.

- The tamper detection phase should be followed by tamper recovery in order to reduce complexity.

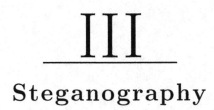

III

Steganography

III

Steganography

Steganography

CONTENTS

THE word steganography consists of two Greek words *steganos* which means covered and *graphia* which means writing. Steganography is nothing but the

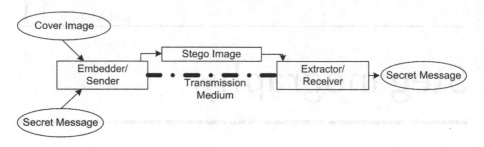

FIGURE 11.1 Basic model for steganography.

concealed communication. When we secretly transfer a message using any carrier/-cover, this phenomenon is called steganography. The carrier/cover and message need not to be in digital form. They may be of any form. In ancient times, there was no availability of digital media and devices, but at that time also, covert communication took place. With the growth of technology, secret communication became more accessible and important. We will discuss the concept of steganography in this chapter in more detail.

The **Chapter Learning Outcomes(CLO)** of this chapter are given below. After reading this chapter, readers will be able to:

1. Understand the basics of steganography.

2. Understand the scope of research in the field of steganography.

3. Develop his/her own efficient algorithm for a steganography approach.

4. Develop the MATLAB code for the basic implementation of approaches.

11.1 INTRODUCTION

Steganography now is known to be a technique for hiding a secret message into a digital signal/image (cover) which can later be extracted or detected. Before studying anything about steganography, we must learn all the basic components and terminology related to it throughout the book. Figure 11.1 shows the basic model of a steganography approach. Various components used in this model are as follows:

1. **Cover or host work:** Cover or host works are those digital signals or images which are used to hide the secret message. The cover work may be either an image or audio or video. Since in this book, we only deal with digital images, hence the default meaning of cover/ host work is digital image. A cover image may be binary, gray or color therefore the complexity and performance of the steganography approach depends on the type as well as dimension of the cover image.

2. **Message/secret information:** A secret message is the information that must be hidden in the cover image. The secret message may or may not be related to the cover image in which it is going to be hidden, but preferably it must be unrelated and this is the main difference between steganography and watermarking.

3. **Embedding system/sender:** The sender or embedding system is the one which embeds the secret message into the cover image. All proposed steganography approaches are executed by this(sender/embedder) entity only. The execution time of the embedding system depends upon the steganography algorithm and the nature of the cover and secret messages.

4. **Stego image:** Stego images are nothing but the cover image embedded with the secret message.

5. **Transmission channel:** It is the medium by which the stego image travels up to the receiver end.

6. **Extraction system/receiver:** This is the entity where the secret message extraction algorithm is executed to extract the secret.

The aforementioned entities are the basic components of any steganography scheme. Each component is related to the others directly or indirectly. A secret message and cover image are both separate entities and both are unrelated mostly. Secret messages and cover images are both treated as input to the embedding system or sender. The embedding system executes the secret embedding algorithm and outputs the stego image. This stego image is transmitted via any secure transmission medium to the receiver. The extraction system separates the secret message from the stego image.

Points worth remembering:
Cover image, secret message, sender and receiver are the main components of any steganography approach.

11.1.1   Watermarking vs. Steganography

Till now we have studied two major image based security techniques: digital image watermarking and visual cryptography. The third technique is image based steganography. Some readers always think that steganography and watermarking are the same but there is a conceptual difference between both. We can understand this difference by an analogy. Let us consider a scenario where a master named "A" writes a secret message on the scalp of his slave and this message must be read by another master named "B". In order to make this message secure, the slave will grow his hair back. Now the slave will travel to master B, where master B again shaves his head to reveal the message. Now here we need to understand the case which will better clarify the difference between steganography and watermarking. Suppose on the scalp of the slave, it is written that "This slave is sent by A now this is yours". It means that the secret message is related to the slave. Here, slave acts as carrier/cover and the message written on the scalp of the slave is the secret message. In this scenario, the secret message is directly related to the carrier/cover. This is the same as the watermarking concept where the embedded secret (watermark) is directly related to the cover. Here the secret message is used just to inform and authenticate the slave. On the other hand, suppose on the scalp of the slave,

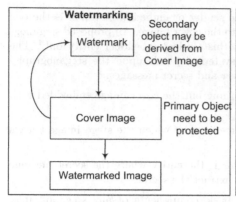
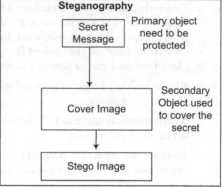

FIGURE 11.2 Comparison between watermarking and steganography.

it is written that "I will come tomorrow" which has nothing to do with slave. This scenario is nothing but steganography where the secret message is unrelated to the carrier/cover. Here carrier/cover is used only to secretly transmit the message. Once the message has reached the receiver, there is no need of carrier/cover. In simple words we can say that in watermarking, the cover is a primary object and the secret message (watermark) is secondary object which is actually used to protect the cover. Whereas in steganography, the secret message is the primary object and the cover is secondary which is used only to cover/protect the secret message. The diference between watermarking and steganography will be more clear from Figure 11.2. Before the evolution of the digital era, there were some other traditional methods of steganography like writing with invisible ink, message transmission by birds etc which were used frequently to secretly share messages.

Points worth remembering:
In watermarking, the cover image is primary and the watermark is a secondary object, whereas in steganography the cover image is secondary and the secret message is the primary object.

11.1.2 Need for Steganography

Now, digital communication increases day by day. Because of its propensity for man in the middle attacks, digital communication is more susceptible to eavesdropping. Actually by using cryptography, messages may be made secure enough or confidential so that no one can understand them. In the case of cryptography, we just encrypt the plain message into the cipher message which is nothing but the secure version of the message. Because of its non-readable nature, the message will be confidential. Idea

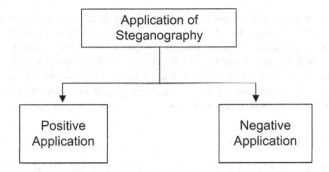

FIGURE 11.3 Categorization of steganography applications.

of cryptography would not work where encryption is illegal as the cipher message might attract attention of an attacker. In such scenario, steganography is the best approach for covert communication without encryption. Here, the message will be hidden behind the cover which is by default confidential and unlike cryptography, due to obvious view of cover, it will not attract the attention of the attacker.

11.2 APPLICATIONS OF STEGANOGRAPHY

As we know, the main objective of the steganography is covert communication so steganography can be used in any place where secret transmission of message is required. We can classify the applications of steganography into two ways: positive applications and negative applications as shown in Figure 11.3. Actually, this classification is subjective because sometimes a negative application may be positive for others or vice versa. For example, communication between two thieves for looting someone may be in the favour of their individual benefits. So, this communication is positive application for them but ethically this is negative application. Here we classify the applications according to one's moral and ethical values.

11.2.1 Positive Applications

Steganography can be used in various places in which secret communications are actually required for the positive welfare or benefits of the country, society or individuals like military application, government secret transmission, bank transactions, lover communication etc. For example Military applications where secure transmission of the messages are highly required. As we know that there are many hackers or attackers of rival countries, which always keep their eyes on the main communication path to steal the secret information. In that case if we transfer secret information which is encrypted using any cryptographic algorithm then this message will not be obvious any more. This phenomenon is called cryptanalysis, where attacker can guess that the given message is not a normal message and there is something important hidden inside it. After analysing this fact, he will try to decrypt the message anyhow. Most of the times cryptography algorithms are available in the public domain. In this case, security of the message is only based on the strength, length and complexity of the secret key. An attacker will always try a brute force attack in order to decrypt the message. Unfortunately it may be possible that someday he will succeed in decrypting the message. What will happen if he is unable to guess the suspicious message using cryptanalysis? This will only happen when all transmitted messages look obvious and normal without any encryption. This objective can be achieved by steganography where all messages are kept behind the covers which always look normal. That is why we can say that steganography is an essential requirement in the aforementioned sensitive applications.

11.2.2   Negative Applications

All applications where steganography is used by criminals or attackers and also ethically against a country, society or individual are called negative applications. There are many countries where political dissent is not tolerated, in that case, dissent organizations may use the steganography techniques in order to communicate with others.

Points worth remembering:
There are two major classifications of steganography, i.e., positive and negative application.

11.3   PROPERTIES OF STEGANOGRAPHY

Just like watermarking techniques, there are eight properties for steganography as shown in Figure 11.4 which must be satisfied up to a satisfactory level. We will discuss these properties one by one.

11.3.1   Fidelity

As we know that in steganography, the cover work and the secret message are totally unrelated and do not depend on each other. Hence we can choose any cover work to hide the secret message. We are free to choose a type of cover image which has less effect after embedding of the secret message. For example if we have a cover image in which the prominent color is blue, because of embedding the secret, it may be possible that the prominent color changes to sky blue. This is not a matter to worry about because our main objective is to protect the secret not the cover.

11.3.2   Embedding Capacity

Embedding capacity is just like watermarking payload. It is nothing but the number of bits of the secret image which can be embedded into the cover work. So the embedding capacity must be as high as possible. We can understand this from the following scenario. Suppose we have a binary image of size $M \times N$ where each intensity is represented by a single bit (either 1 or 0). In this scenario, we do not have a choice of selecting the bits to embed the secret message. This scenario may be

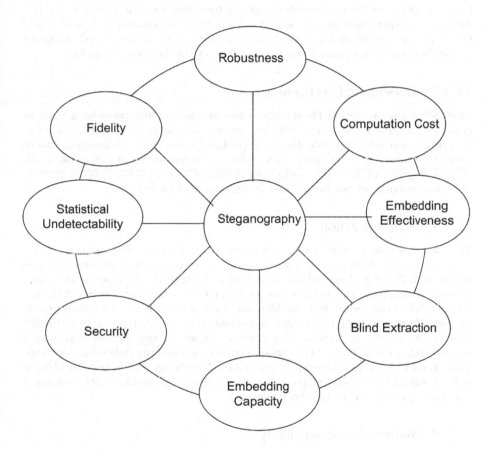

FIGURE 11.4 Various properties of steganography.

a good example for less embedding capacity. Embedding capacity can be enhanced just by choosing the right cover image. Suppose we choose a gray scale cover image of size $M \times N$. In this case we have total $M \times N \times 8$ number of bits among which we can choose the sufficient number of bits to embed the secret message so that the fidelity property is also maintained. For better imperceptibility of the cover image after embedding the relation, $M \times N \times 8 >> M \times N \times k$ must be satisfied where k is the number of bits per pixel dedicated to embedding the secret message. The embedding capacity can be enhanced significantly when we choose a color cover image. In this case the total number of bits in the entire image is $M \times N \times 24$ (eight bits per pixel per color plane). Now we have a sufficient number of bits to embed the secret message maintaining the fidelity property. Here we need to pay attention to the dominant color plane so that embedding affects the cover image less.

11.3.3 Embedding Effectiveness

Embedding effectiveness is the quality of the embedder. Our embedder is must be capable enough of embedding the secret message into the cover work in such a way that no steganalysis is possible. Here we introduced a new word "steganalysis" which is similar in meaning to "cryptanalysis". The word steganalysis consists of two words "Stegano" and "Analysis". It means the presence of an embedded secret message in the cover image must not be deduced by an attacker by any means.

11.3.4 Blind Extraction

This case is similar to blind watermarking where we do not need to send an original cover image with the embedded one. Actually here we assume that the receiver knows all the embedding and extraction algorithms along with the locations of the secret message embedding. This is the safest method for steganography because if there is a plain cover image with the embedded one, then one can know the message easily by just subtracting both the images. Sometimes this blind extraction is also called informed extraction because at that moment the sender and receiver both have a common set of cover images. They select a cover image and the embedding locations. This scenario is simpler because at this time, the receiver will have both plain as well as embedded versions of the cover image. Hence extraction of the message is very easy (just by subtracting both versions).

11.3.5 Statistical Undetectability

Actually the main objective of steganography is to hide the basic fact that a secret communication is happening between the sender and receiver. There are many statistical methods such as histogram analysis which are used in the process of steganalysis to see a specific pattern of communication. Our steganography algorithm must have the statistical undetectability feature.

11.3.6 Robustness

Since secret messages are hidden behind the cover images, embedding must be robust enough so that it can tolerate intentional or unintentional attacks. Here we assume that the embedded cover image (also called a stego image) travels through a digital network. A digital network is more vulnerable to man in the middle attack. Hence our stego image must be robust enough to tolerate any pixel or signal-based modifications. It means that if the stego image is modified in any way it still must provide the exact hidden secret message.

11.3.7 Security

As we know that algorithms are always well known and also publicly available. Hence if an attacker knows the algorithm for embedding, the secret message can easily be extracted. Hence the embedded message must be secure enough so that after extraction, no one can interpret it. This can only be made possible when the secret message is encrypted by a symmetric key before embedding. This secret key will be known only to the sender and receiver, so at the time of extraction, unlike the attacker, the receiver can only decrypt the message using the same key.

11.3.8 Computation Cost

The computational cost of any steganography approach should be as minimum as possible. There is a trade-off between the computational cost and security of the steganography approach. If we reduce the cost of the secret message embedding techniques such as embedding the secret message in very straightforward manner without much complexity, then we are definitely somewhere compromising the security because this type of message will be easier to extract by an attacker. Similarly if we increase the computational cost, then it definitely will return a good and secure steganography scheme. We need to develop a type of approach that is balanced.

Points worth remembering:
There is a trade-off between computational cost and security in steganography.

Points worth remembering:
There are eight properties of steganography. They are robustness, embedding effectiveness, computation cost, embedding capacity, security, fidelity, statistical undetectability and blind extraction.

11.4 PERFORMANCE MEASURES FOR STEGANOGRAPHY APPROACHES

If there is more than one steganography method then selecting the best one is very important and a difficult task. There is no fixed set of performance parameters

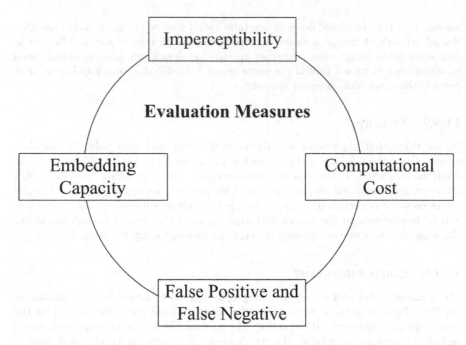

FIGURE 11.5 Various evaluation measures for steganography.

available to evaluate and compare two or more steganography methods, but we can still discuss a few parameters by which we can compare the algorithms as shown in Figure 11.5.

11.4.1 Embedding Capacity

Embedding capacity is a quantitative parameter and can be measured in terms of bits. That is why it can be considered a good evaluation measure. A good steganography approach must have a high embedding capacity. Embedding capacity highly depends on the type of cover image. If we choose a binary image as cover, then the embedding capacity will be very low whereas, color or multidimensional images have a very high embedding capacity. Similarly the image content also plays a very important role for embedding capacity. For example if we have a very smooth cover image with average intensities, then embedding of the big secret message in it may be vulnerable to steganalysis. Similarly if we take a cover image with random and different intensities, then after embedding of a large number of bits of the secret, one cannot decide the presence of any other information so it is a challenge for steganalysis. Unlike watermarking, it is up to the sender or embedder which image must be chosen as cover because the secret and cover are unrelated to each other.

11.4.2 Imperceptibility

Whenever we embed the secret message into a cover image, then we must ensure a high imperceptibility between the cover and stego image. This is also one of the parameters for evaluating and comparing two or more steganography approaches. We already discussed how imperceptibility can be measured by PSNR, SNR, MSE, etc., in previous chapters.

11.4.3 False Positive and False Negative

Actually the secret message is embedded into the bit level depth of the cover image therefore we need to compare the plain cover and stego image using bit level depth. In order to have the smallest differences between the cover and stego image, we need to minimize the false positive and false negative rate.

11.4.4 Computation Cost

Two or more steganography methods may also be compared on the basis of computation cost. Computational cost may be measured in terms of time complexity, computational complexity or space complexity. A good steganography approach must have a proper balance among all. For example if we created an algorithm which is secure enough against any man in the middle attack or steganalysis, but extraction of secret takes too long, then this approach will be undesirable, because sometimes we need fast recovery of the secret with some level of security.

Points worth remembering:
There are various parameters such as computational cost, embedding capacity, imperceptibility etc with which two or more steganography approaches can be compared.

11.5 MATHEMATICAL NOTATION AND TERMINOLOGY

In this section we define steganography mathematically. Let K, M and C be sets of secret keys, secret messages and cover images, respectively. Let K_i, M_j and C_k denote the elements from the set K, M and C, respectively, where i, j, k may or may not be equal. A steganography scheme can be understood with the help of two functions: embedding E_m and extraction E_x which can be represented as

$$E_m : C \times M \times K \to S \tag{11.1}$$

where S is the set of stego images. Here $\times$ shows that we can choose any of the cover images from set C and any secret keys from set K in order to hide any of the message from set M. For instance, suppose we are taking unique elements from all the sets then the relation will be,

$$E_m : C_k \times M_j \times K_i \to S_l \tag{11.2}$$

Now the extraction function E_x can be defined as

$$E_x : (S_l, K_i) \to M_j \tag{11.3}$$

Here the same secret key (as used at the time of embedding) must be used in order to extract the orginal hidden, message from the stego image. The relation between function E_m and E_x can denoted as

$$E_x(E_m(C, K, M), K) \to M \tag{11.4}$$

Here the cardinality $|M_j|$ for any message shows the number of bits which are to be hidden and cardinality $|C_k|$ shows the number of bits in which M_j can be embedded. That is why $|M_j| << |C_k|$ and $\frac{|C_k|}{|M_j|}$ shows the number of embedding vacancies for a single bit of message into the cover image. The beauty and efficiency of our algorithm is in how we utilize these vacancies or spaces for embedding. The stego image S for any cover image C is always a little bit distorted version of C. This distortion must be as small as possible and it can be measured by $D(S, C)$ where D is any function which can be used to measure the difference between C and S.

Points worth remembering:
Any steganography approach can be defined by only two functions embedding E_m and extraction E_x.

11.6 STEGANALYSIS

A method which is used to detect the presence of any hidden secret message in the stego image is called steganalysis. There are three types of steganalysis possible: passive, active and malicious as shown in Figure 11.6. Each one has its own role for detecting and altering the the stego image in such a way that normal communication between sender and receiver gets interrupted.

11.6.1 Passive Steganalysis

In this case the attacker tries to intercept the presence of the message in the stego image. If he does not find any message, then he simply forwards the stego image to the receiver. In this case, if he detects a message inside the stego image, then he just blocks the transmission to the receiver. Hence receiver would not be able to get any communicated message from the sender side which contains the secret message.

11.6.2 Active Steganalysis

In this case the attacker has more freedom. He can even modify the content of the communication between the sender and receiver. If the sender and receiver are

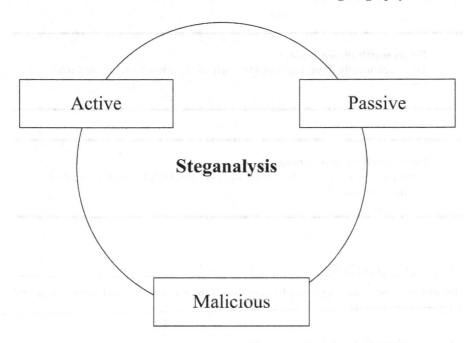

FIGURE 11.6 Various types of steganalysis.

normally communicating with each other and the attacker just senses that a secret is also being transmitted behind the normal communication, then instead of knowing the secret message, he will try to tamper with the stego, so that the hidden secret will also be tampered with, if it is not robust enough. For example if the sender and receiver are communicating through the transmission of normal images, then the attacker compresses the transmitted image in between the communication. If the stego image is not robust against a compression attack, then the embedded secret message will also be tampered and it will be meaningless at the receiver side,

11.6.3 Malicious Steganalysis

This is the most dangerous type of steganalysis because it is one step ahead of both passive and active steganalysis. Here we assume that the attacker cannot only sense the presence of a secret message in the stego image, but also understands its meaning. This is only possible when the attacker knows the steganographic algorithm with the chosen secret key. In this scenario, the attacker can impersonate the sender and sends the wrong message to the receiver on behalf of the sender. This is a higher level of analysis where besides the detection of the presence of the secret message, attacker also tries to interpret and change the secret message. This method can also be called *forensic steganalysis*.

Points worth remembering:
There are mainly three types of steganalysis viz passive, active and malicious.

Points worth remembering:
Malicious steganalysis is the most dangerous steganalysis and is also called forensic steganalysis.

11.7 DETECTION

Detection of secret messages can be classified into two classes: blind steganalysis and targeted steganalysis.

11.7.1 Blind Steganalysis

In this class we assume that the attacker has some level of suspicion that the sender and receiver are covertly communicating. But the attacker does not know which steganography algorithm is used during communication. So here the attacker must develop a type of steganalysis algorithm that is capable of detecting a wide range of steganography algorithms.

11.7.2 Targeted Steganalysis

If an attacker can guess the steganography algorithm used during covert communication, then he will create a type of steganalysis algorithm that detects just that specific class of steganography. This type of detection algorithm is called targeted steganalysis.

In both classes, an attacker must be versed in pattern recognition, machine learning, image processing, etc. It is impossible to know the probability distribution for a whole set of cover images as well as stego images available in the space. Hence we cannot construct an optimal classifier to test their likelihood ratio. The efficiency of a classifier can be tested by two errors i.e., false acceptance and false rejection. If a cover image is treated as a stego image by a classifier, then it is called false acceptance and if a stego image is treated as a cover image, then it is called false rejection. Any image is represented by various features so in steganalysis our objective is to extract one or more features of the image which are used to construct the image. The choice of the features is crucial because if we choose good features, then we can create an accurate classifier, otherwise it is a poor classifier. Ideally feature vector of all the features of the cover image and its corresponding stego image will never overlap with each other. In this case classification is very easy but in practice, this is not the case because some overlapping is present in the boundary regions of both feature vectors.

This overlap is only responsible for both types of classification errors (false rejection and false acceptance).

Points worth remembering:
A steganalysis scheme which is created for all classes of steganography algorithms is called blind steganography, and a steganalysis scheme created for a particular class of steganography is called targeted steganalysis.

Some staganalysis methods run in the reverse direction. They first estimate the cover image from the stego image, after that just compare the cover and stego to find the difference and that difference is nothing but the secret message. Usually it is very difficult to find the exact cover image from the stego but the some steganalysis methods can work on an approximation of cover work only. The estimation process from the stego to cover work is called calibration.

SUMMARY

- The *cover image, secret message, sender and receiver* are the main components of a steganography approach.

- In watermarking, cover image is primary and the watermark is secondary object whereas in steganography, the cover image is secondary and the secret message is the primary object.

- There are two major classifications of steganography i.e. *positive and negative applications*.

- There is a trade-off between *computational cost and security* in steganography.

- There are eight properties of steganography which are *robustness, embedding effectiveness, computation cost, embedding capacity, security, fidelity, statistical undetectability and blind extraction*.

- There are various parameters such as *computational cost, embedding capacity, imperceptibility etc* with which two or more steganography approaches can be compared.

- Any steganography approach can be defined by only two functions, embedding E_m and extraction E_x.

- There are mainly three types of *steganalysis, namely, passive, active and malicious*.

- Malicious steganalysis is the most dangerous type of steganalysis and it is also called *forensic steganalysis*.

- A steganalysis scheme which is designed for all classes of steganography algorithms is called *blind steganography* and a steganalysis scheme designed for a particular class of steganography is called *targeted steganalysis*.

CHAPTER 12

Development of Steganography and Steganalysis

CONTENTS

I N the previous chapter we covered some basic definitions and concepts of steganography and steganalysis. We saw various properties and evaluation parameters for comparing the efficiency of two or more steganography methods. Just like watermarking there are various development methods, available to generate an efficient steganography method. In this chapter we will see some basic development models for steganography as well as steganalysis methods. Readers can develop

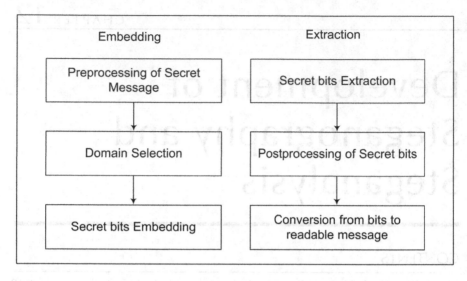

FIGURE 12.1 Phases of steganography.

their own methods after understanding of basics given in this chapter. The **Chapter Learning Outcomes(CLO)** of this chapter are given below. After reading this chapter, readers will be able to:

1. Understand the basic development of a steganography approach.
2. Understand the basic development of a steganalysis approach.
3. Develop his/her own efficient algorithm for a steganography and steganalysis.
4. Develop the MATLAB code for the basic implementation of the approaches.

12.1 PRACTICAL APPROACH TOWARDS STEGANOGRAPHY

Development of any steganography method is done through two processes: embedding and extraction of the secret message as shown in Figure 12.1. Here we can see that both processes are further subdivided into three steps. These are the major steps which must be taken into consideration during the development of steganography methods. First of all we will deal with an embedding approach for a secret message.

Points worth remembering:
Steganography methods are divided into embedding and extraction processes.

12.2   EMBEDDING OF A SECRET MESSAGE

In this section we will go through some of the techniques of embedding of secret messages. As we have seen, there are three steps in embedding: preprocessing, domain selection and bit embedding. One can make his/her own customized approach using these steps. Actually this is a hierarchical approach, one can modify it in order to reduce the complexity of the algorithm. The embedding approach is most similar to the watermarking embedding methods with the major difference of isolation of secret and cover. In steganography, there is no dependency between the secret message and cover image. Here we are least concerned with the quality of the stego image after embedding. Our main objective is embedding the secret message with high capacity and security. We need to make a type of approach that satisfies all the essential properties of steganography.

Points worth remembering:
There are three steps of embedding: Preprocessing, Domain selection and Bits embedding.

12.2.1   Preprocessing a Secret Message

There are basically three steps for preprocessing a the secret image in steganography as shown in Figure 12.2. These three steps are: lossless compression, ASCII code conversion and binary conversion. The circular direction shows the freedom of any preprocessing algorithm starting from any step. As we know that ultimately only bits 0 and 1 can be inserted into the bit depth of the cover images, so we need to convert the secret messages into the bit stream of 0 and 1 in any preprocessing algorithm.

This preprocessing step is used to enhance the embedding capacity of steganography. In order to increase the embedding capacity we have two options:

1. Take the cover image with high bit depth (color or multi-spectral images).

2. Compress the secret message.

Taking the high bit depth image is very costly. Hence we need to compress the secret message before proceeding towards the development of any steganography method. We can say that this is the preprocessing of secret message before embedding.

Points worth remembering:
There are three sub-steps in preprocessing which are lossless compression, ASCII code conversion and binary conversion.

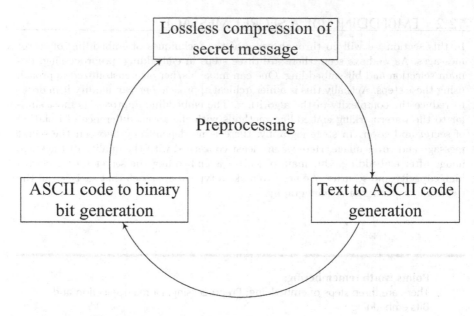

FIGURE 12.2 Various steps in preprocessing.

12.2.1.1 Lossless compression of secret message

The secret message is encoded into its compressed version before embedding into the cover image. As we already know that compression methods can be broadly classified into two types: lossy compression and lossless compression as shown in Figure 12.3. In lossy compression methods, we cannot exactly recover the original data after decompression. They are generally used for image and video compression where some loss of information can be ignored by our human visual system. But this loss of data is not desirable if our target data is textual information, because a change in any letter can change the entire meaning of the secret. Hence in the case of textual data, which is the general form of a secret message in the case of steganography, we use lossless compression methods. Thus, at the receiver end, we exactly recover the same information as it was before compression. Lossless compression of a secret message may be divided into two types: fixed length encoding and variable length encoding.

Points worth remembering:
Lossless compression of a secret message is divided into two types: fixed length encoding and variable length encoding.

Fixed length coding:
When we have a secret message that is full of redundant characters, then a fixed

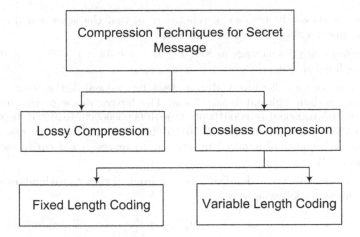

FIGURE 12.3 Classification of compression techniques.

length coding method will be the most suitable compression method for that secret message. Here we need to follow a few steps in order to encode the secret message:

1. First of all take the secret message as input to the encoding system. Let's take an example of a secret message: "Yaaaaaa Bllaaaaaahhhh". Here we see that many characters like a and h are continuously repeated throughout the message.

2. We need to replace the characters with their ASCII values. In this case the stream of ASCII values will be "89 97 97 97 97 97 97 32 66 108 108 97 97 97 97 97 97 104 104 104 104". Here we can see that all characters including space are replaced with their ASCII values. The main thing to remember here is that the characters from a to z or A to Z and numbers from 0 to 9 with some special characters (like space) can be represented only by the the ASCII values from 0 to 255. It means the ASCII value of every character of the secret message can be represented in eight-bit binary form.

3. If we convert the aforementioned secret message into its binary form, then the binary bit stream will be "0101100101100001011000010110000101100001011000010110 0001011000100100000010000100110110001101100011000010110000101100001011000 0101100001011000010110100001101000011010000110100001101000". Here each character is represented by eight bits (fixed) of binary representation. There are $21 \times 8 = 168$ binary bits in this message. Since we already know that this message can be encoded by fixed length encoding in order to increase the embedding capacity, we will see how this idea can be implemented.

4. The message will be written in the following manner:
Eight bit binary (ASCII of character) $\times$ eight bit binary (repetition of the character) || eight bit binary (ASCII of character)$\times$ eight bit binary (repetition of the character) and so on

Here $\times$ shows the number of repetitions of that character and $\|$ represents the concatenation of the next character.

5. We can write the message as $89 \times 1 \| 97 \times 6 \| 32 \times 1 \| 66 \times 1 \| 108 \times 2 \| 97 \times 6 \| 104 \times 4$.

6. Now again we will convert all the characters as well as the number of their instances into eight-bit binary form. The binary vector of this message is "0101100100000001011000010000011000100000000000010100001000000001011 0110000000010011000010000011001101000000000100". Here we can see that now the number of required binary bits to represent the entire secret message is 112.

7. Let's calculate the redundancy in the original secret. Redundancy can be calculated in following way:

$$R_d = 1 - \frac{1}{C_r} \tag{12.1}$$

where C_r is the compression ratio and calculated as:

$$C_r = \frac{N_1}{N_2} \tag{12.2}$$

where N_1 and N_2 are the number of bits in the original and compressed messages, respectively.

8. In the given example $N_1 = 168$ and $N_2 = 112$, then $C_r = \frac{168}{112} = 1.5$. Hence Redundancy $R_d = 1 - \frac{1}{1.5} = 0.33$ or we can say that in the original secret message, there was 33% redundancy.

9. Once we get the compressed binary coding of the secret message, we need to just embed it into the cover image. We will see the process of embedding in detail in the next section.

Points worth remembering:
In fixed length coding, all the characters of secret message are encoded into the same length bit stream.

Variable length coding:
Variable length coding can be used in both cases viz when we have continuous repetition of the same character as well as when we have a message with different non-continuously repeated characters in order to compress the secret message. We will see this compression method in an example given below:

1. Let us consider a secret message S which consists of six unique characters. To apply variable length coding, we need to calculate the probability of occurrence for each character. The probability can be calculated as:

$$P_i = \frac{n_i}{N} \tag{12.3}$$

where P_i is the probability of occurrence of the ith character where the $i = 1..6$ for this example, n_i is the total number of occurrences of ith character and N is the size of the secret message (total number of characters in it). For example in word the "apple" the probability of "p" is $\frac{2}{5}$.

2. Figures 12.4 and 12.5 are the examples to explain the concept of variable length coding. Here we see that first we calculate the cumulative probabilities of all characters and they are arranged in non-increasing order in the new column. This is repeated till there are only two entries left in the column. In this example there are six characters denoted by C1, C2 and so on. After adding all the probabilities finally, we find two cumulative probabilities which are 0.6 and 0.4. Now we will go in the reverse direction for assigning the bits to each probability as shown in Figure 12.5. Finally we see that each character is assigned a unique variable length code.

3. The main thing to observe is that the character with the highest probability will be assigned the least number of bit representation and similarly no assigned bit stream to any character will be prefixed to any other bit stream of a character. For example character C2 is assigned bit stream 00, so coding of no characters will start from 00.

4. Now we can calculate the average number of bits per pixel as follows:

$$L_{avg} = \sum P_i \times b_i \tag{12.4}$$

where P_i is the probability of the ith character and b_i is the number of bits assigned to that character. Now we can calculate the L_{avg} for this example as

$$L_{avg} = 0.4 \times 1 + 0.3 \times 2 + 0.1 \times 3 + 0.1 \times 4 + 0.06 \times 5 + 0.04 \times 5 = 2.2 \tag{12.5}$$

It means that an average 2.2 bits can be assigned per pixel. Without any compression eight bits per pixel were assigned.

5. One can decode the secret message by moving from left to right of the encrypted bit streams. For example if a compressed bit stream is 011001, then in order to decompress it, we move from left to right and as soon as we find a code for some character we use that for decryption. Start reading 0, as this is not assigned to any character, then read 01 which is also not assigned to any character, then read 011 as this sub bit stream is assigned to C3. Now again start from the next 0 bit and so on. Finally we get the decompressed characters as C3C2C1.

Here we have just shown the examples for run length coding for fixed length and Huffman coding for variable length coding. This is not a hard and fast rule to apply to only these compression methods. There are many other lossless compression methods available which can be used with better efficiency. Readers can design their own lossless compression approach as per their requirements for steganography.

Points worth remembering:
In variable length coding, all the characters of the secret message are encoded into different lengths of bit stream.

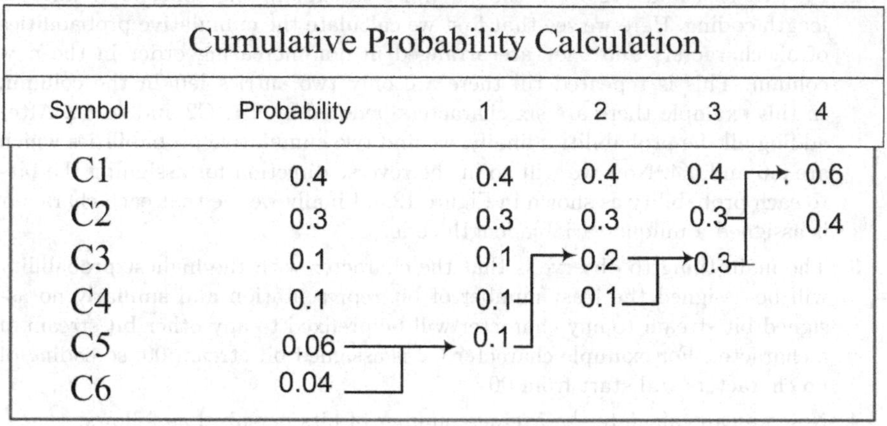

Cumulative Probability Calculation					
Symbol	Probability	1	2	3	4
C1	0.4	0.4	0.4	0.4	0.6
C2	0.3	0.3	0.3	0.3	0.4
C3	0.1	0.1	0.2	0.3	
C4	0.1	0.1	0.1		
C5	0.06	0.1			
C6	0.04				

FIGURE 12.4 Cumulative probability calculation for the secret message.

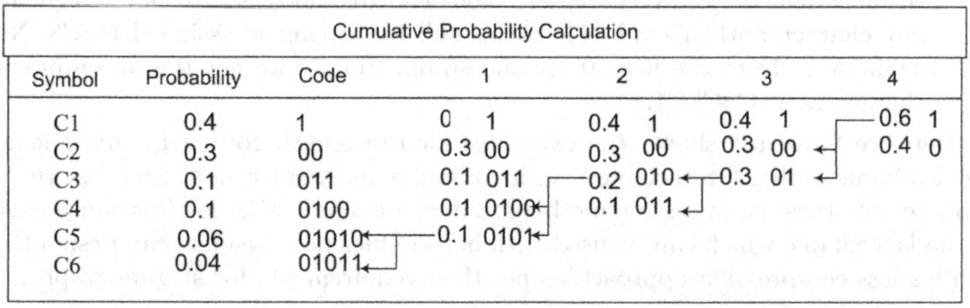

Cumulative Probability Calculation						
Symbol	Probability	Code	1	2	3	4
C1	0.4	1	0 1	0.4 1	0.4 1	0.6 1
C2	0.3	00	0.3 00	0.3 00	0.3 00	0.4 0
C3	0.1	011	0.1 011	0.2 010	0.3 01	
C4	0.1	0100	0.1 0100	0.1 011		
C5	0.06	01010	0.1 0101			
C6	0.04	01011				

FIGURE 12.5 Bits assignment to each character of the secret message.

12.2.1.2 Text to ASCII code generation

In the aforementioned compressions, we saw the that the compressed text is converted into ASCII format. Here we will see the significance of this step. Actually text to ASCII conversion can be done before or after the lossless compression. There is no strict rule; it is just up to the developer's discretion. ASCII code is a well-known code hence the sender and receiver can agree on that. But they can design their own text to number mapping code. By this way they can enhance the security also. In ASCII code we assume that we need at max eight bits per character in order to encode that. It means we can have a total of 256 characters including all characters, numeric and special characters. There may be many cases where a secret message may have far fewer number of characters. In such cases, ASCII code may not be a better option. For example if we have a secret message which has only fourteen different characters and numbers, then we can denote it in binary form by using only four bits as $2^4 = 16$. If we use ASCII code for this, then unnecessarily we will have to use four more bits for each character. User-defined coding is also beneficial when we talk about a security point of view because ASCII codes are well known and if an intruder understands the embedding algorithm, then he can easily decode the secret message. Decoding a secret is more difficult when we use user-defined code.

Points worth remembering:
User-defined text to number mapping is more desirable for security as well as an embedding capacity point of view.

12.2.1.3 ASCII code to binary bit conversion

As we know that ultimately only bits 0 and 1 will be inserted into the bit planes of the cover images so we need to convert the ASCII code of the secret messages into the bit stream of 0 and 1 in preprocessing. When we have a secret message in the form of streams of bits, then it is up to us how we insert it into the cover image. In order to make a steganography approach more secure, we can shuffle the bits of the secret message using any symmetric key. At the receiver end, the same key will be used to decrypt the encrypted bit stream.

Points worth remembering:
Shuffling a secret bit stream is desired to provide confidentiality even after extraction of the secret bit stream.

12.2.2 Domain Selection

After getting the binary bit stream for the secret message, we need to embed it into the cover image. We have two domain options in the cover image for embedding: the first one is the spatial domain and the second one is the frequency domain. Both domains have their own pros and cons, so we need to select the appropriate domain as per our requirements. First, the basic pros and cons associated with each domain are explained. First the preprocessing step deals with the property of steganography that is the embedding capacity. But here we deal with almost all the properties of steganography like robustness, undetectability, blind extraction, security, etc. Selection of the appropriate domain is responsible for achieving the desired ideal values of the properties.

12.2.2.1 Spatial domain

When we deal directly with pixel value in the cover image, then that domain is called the spatial domain. The benefits and drawbacks of this domain are as follows:

Benefits of the spatial domain:

1. Processing with the spatial domain is less complex and so its computational cost is lower.
2. One can easily regulate the embedding capacity in the case of the spatial domain. Since the quality of the stego image is not the concern, one can very easily modify the embedding capacity.
3. One can achieve the blind extraction benefits when embedding in the spatial domain.

Drawbacks of the spatial domain:

1. Embedding in the spatial domain is statistically less undetectable. Using steganalysis one can easily detect the existence of the secret message if the embedding algorithm is not secure enough.
2. It is less robust against various intentional or unintentional attacks. It means the secret message can be easily destroyed after any attack if designed algorithm is not efficient enough.
3. It is less secure because using a brute force attack, one can decode the secret message once extracted.

12.2.2.2 Frequency domain

When we deal with the frequency coefficients of the pixel values of the cover image, then that domain is called the frequency domain. Frequency coefficients of the cover image can be calculated by a number of methods such as discrete cosine transform, discrete fourier transform etc. The benefits and drawbacks of this domain are as follows:

Benefits of the frequency domain:

1. Embedding in the frequency domain is statistically more undetectable. It is very difficult in steganalysis to detect the existence of a secret message if embedded into the frequency domain.

2. It is more robust against various intentional or unintentional attacks. It means the secret message cannot be easily destroyed after an attack if the secret message is embedded into the frequency domain.

3. One can achieve the blind extraction benefits when embedding in the spatial domain.

Drawbacks of the frequency domain:

1. One cannot easily regulate the embedding capacity of the secret message in the case of the frequency domain.

2. Processing with the frequency domain is more complex and its computational cost is higher.

Points worth remembering:
Both domains have their own pros and cons, so we must select the one that fits our requirements.

12.2.3   Secret Bits Embedding

So far, we have covered one method for generation of secret bits from the secret message and the significance of domains (spatial and frequency). Now our objective is to develop a method to embed the secret bits into appropriate selected domain. There are many methods, but here we will discuss one for each domain.

12.2.3.1   Embedding in the spatial domain

We know that in the spatial domain, we deal directly with pixel intensities. Hence here we need to embed the secret bits into the bit planes of the cover image's pixel. In the earlier chapter on digital watermarking, we saw much important information related to watermark embedding in the spatial domain. Embedding of secret messages in the case of steganography is just like watermarking, but the main difference here is, we do not use a self-embedding approach. In self-embedding, the watermark bits are generated from the cover image itself, but here there is no case like this. In steganography, we already have our secret bits to be embedded and we just need a good embedding mechanism to embed them into the bit planes of the cover image.

Embedding in all pixels:

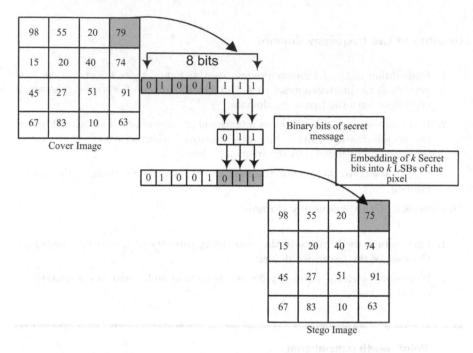

FIGURE 12.6 Example of steganography in spatial domain.

As we know that the LSBs of pixels are the best place for embedding because they do not deteriorate the quality of the cover image too much. As we increase the number of LSBs per pixel for embedding, the imperceptibility between the cover and stego images will get reduced. In steganography, the secret message is primary and the cover image is the secondary object, so we are not bothered about the degraded quality of the stego image after embedding. Let us consider a cover image I of size $M \times N$. If we take three LSB bits per pixel to embed the secret message, then the minimum and maximum change in the intensity may be 0 and 7, respectively. This is negligible and has much less effect on the visual quality of the stego image. Here we have in total $M \times N \times 3$ number of vacant locations for embedding the secret message. If the secret message is already compressed, then we can manage with one or two bits per pixel also. We can understand this concept from Figure 12.6, here embedding in only one pixel of the cover image is shown. The same procedure will be repeated for all the pixels of the cover image. Here we can see that three secret bits 011 are embedded into the three LSBs of pixel intensity 79 and because of this embedding, the pixel intensity is modified as 75. The relative difference between the original and modified pixel is 4 which has much less effect visually.

Points worth remembering:
Embedding in all pixels is less secure but there is good embedding capacity.

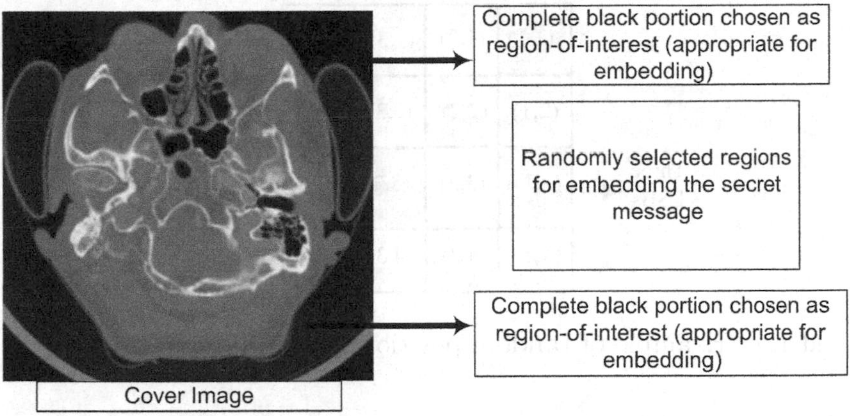

FIGURE 12.7 Example of steganography in spatial domain with region-of-interest (*Courtsey of Dr. Luca Saba, Radiology Department, University of Cagliari, Italy*).

Embedding in Region-of-Interest:

If the number of secret bits is denoted by S_b and $S_b << M \times N \times 3$, then we have many choices to embed the secret bits into the pixels of the cover image. In order to enhance the security and imperceptibility of the stego image we can select some region-of-interest for embedding. The region-of-interest is selected in a very secure manner and in such a way that no attacker can judge the existence of the secret message. We can understand it from Figure 12.7 where we take selected black regions for the embedding of the secret message. These black regions are called a region-of-interest. Embedding locations will not be continuous and the random embedding positions will only be known by the sender and receiver of the message.

Points worth remembering:

Region-of-interest-based embedding is appropriate when we have less but very important secret information to be embedded.

One can also enhance the security by encrypting the secret bits with the help of a secret key. If an attacker attacks the encrypted secret bits, even then he will not be able to decrypt and read the message. The region-of-interest for embedding of the secret message can also be calculated by a user-defined algorithm without referring to the cover image visually. Here we suggest one algorithm to determine the random locations for embedding.

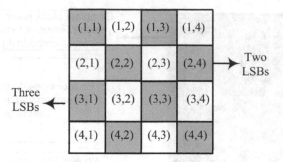

FIGURE 12.8 **Example of random position selection.**

Steps to find random locations for embedding:

1. Let $P(i,j)$ be the pixel intensity of $(i,j)^{th}$ location of cover image I where $1 \leq i \leq M$ and $1 \leq j \leq N$.

2. We will only select those positions for embedding where i and j both will be either even or odd together.

3. To enhance the security one step ahead we can take only two LSBs of $P(i,j)$ for embedding if $i = j = even$ and three LSBs if $i = j = odd$

The aforementioned algorithm is shown in Figure 12.8 where for all selected portions either i and j both are even or both are odd. We are choosing two LSBs from the even positions and three LSBs from the odd positions. This is just an example for randomly selecting the locations for embedding in such a way that blind extraction of the secret is possible at the receiver end. If the receiver knows the algorithm for random selection, then he does not need any embedding locations. He will just search for all even and all odd positions with respect to i and j. After that, he will just extract the two or three LSBs from the pixel intensities as written in the algorithm.

This is one algorithm for steganography in the spatial domain. Readers can create their own effective algorithm with new random location selection mechanisms. They can also create a good compression and embedding algorithm which is less vulnerable to steganalysis.

12.2.3.2 Embedding in the frequency domain

In the previous section, we saw the embedding of secret bits into the spatial domain of the cover image. This approach is good in some cases, but the embedded secret message is not robust at all. To make it robust, we can embed secret bits into the frequency domain. It is just like robust watermark embedding which we have already seen in earlier chapters. In this case, embedding capacity is reduced dramatically in comparison to the spatial domain as here a single secret bit will be embedded into the block of pixels of the cover image. Greater robustness requires large size blocks of the pixels. Here we mention one method for steganography in the frequency domain.

1. First of all transform the cover image using any transformation function like DCT, DFT, etc.

2. Now we need to divide this transformation matrix of the cover image into the non-overlapping blocks of size $m \times n$. One block hides one secret bit which is why if we take a large size block, then the embedding capacity will be reduced automatically.

3. Apply the proposed algorithm on each block in order to embed the single secret bit.

4. Once we embed all the bits in the blocks, then just take the inverse transformation of the matrix. The resultant image is called a stego image.

We can understand the aforementioned algorithm by Figure 12.9. Here we have divided the transformation matrix of the image into the non-overlapping blocks of size 8×8. In this figure we have shown the operation on only a single block. All indexes of the blocks are numbered from 1 to 64. We are taking three pairs of evenly placed indexes from the block which have indices 15, 22, 43, 50, 47 and 54. As per the algorithm shown in the figure, if the current secret bit is 1, then modify the coefficient of indices 15, 43 and 47 in such a way that $15 > 22$, $43 > 50$ and $47 > 50$, otherwise vice versa. At the receiver end just by comparing the coefficients one can judge the hidden bit in that block. This is a very trivial approach for hiding the secret bits in the frequency domain. Readers can create their own efficient approach to achieve ideal values for the maximum properties in steganography.

Points worth remembering:
Embedding in the frequency domain is robust enough because of the reversibility property of the transformation matrix.

12.3   EXTRACTION OF SECRET MESSAGES

So far, we have seen the basic approaches for embedding a secret message into the cover image. This embedded cover image is called a stego image. Now that stego image will be transmitted through a secure channel. Here we assume for a moment that there is no steganalysis or man in the middle attack. In this case, the receiver must be able to extract exactly the same secret message as it was embedded. Just as in the embedding steps, we again follow the hierarchical approach but in the reverse direction for extraction of the secret message.

12.3.1   Secret Bits Extraction

At the receiver end, first we need to extract the embedded secret bits. The extraction process of the bits is subjective because it totally depends on the embedding process. For example if we had chosen the spatial domain for embedding, then its extraction process will be easier. In this case we just need to extract the secret bits from the LSBs of the pixels. This will be a little bit difficult when the secret bits are embedded

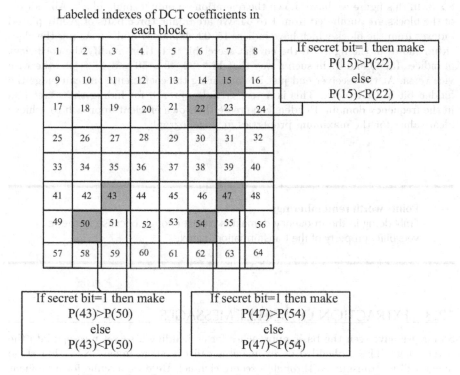

FIGURE 12.9 Example of steganography in frequency domain.

into a region-of-interest rather than in all pixels. The receiver must locate all the embedded pixels using the region-of-interest selection algorithm. After that he must be aware of how many LSBs are dedicated to each ROI pixel. Once everything is clear, then the receiver just extracts the secret bit stream. Here the main thing which he must be careful of is the sequence of the secret bits. Extraction of bits must be in such a way that the sequence of the extracted and embedded bit streams are the same.

If embedding is done in the frequency domain, then extraction will be more complex because the receiver needs to follow the same process as in embedding. He needs to calculate the same transformation in order to calculate the exact embedded secret bits. After the transformation, he must divide the matrix of coefficients into the same number of non-overlapping blocks. Once he extracts the secret bits, then he needs to reshuffle it using the symmetric secret key if they were shuffled at the time of embedding.

Here our assumption is that there must not be a man in the middle attack. What would happen if there were no attack on the stego image? First of all we assume that embedding is done in the spatial domain, as any attack either intentional or unintentional would modify the pixel intensities of the stego image. Since all secret bits are embedded into the LSBs of the pixel intensities, there may be some loss in the secret messages. That is why we say that embedding in the spatial domain is not robust enough. If the same case were to happen with the stego image in which secret bits are embedded into the frequency domain, then because of the reversibility property of the transformation matrix, one could again get the destroyed secret bits.

Points worth remembering:
Extraction of secret bits from the spatial domain is much easier than extraction from the frequency domain.

12.3.2   Postprocessing of Secret Bits

The extracted secret bits from the above step are now further processed. This step is called postprocessing because this is just the reverse of the preprocessing step. Actually the extracted secret bits may be in compressed form. So before making them into textual form we need to decompress them. Otherwise we would not be able to get the exact transferred secret message. Sender and receiver both must have common consent for any compression algorithm, either fixed length or variable length coding. For example, if fixed length coding is used and let's say we have a bit stream "1100010100000100". It means the first eight bits show the binary representation of pixel intensity and the next eight bits show its occurrence. So after decompression, the expanded bit stream will be 1100010111000101110001011000101. If variable length coding is used, then the receiver must be aware of the bit length of the characters. Once the extracted secret bits are decompressed then we need to convert them into readable format.

12.3.3 Conversion from Bits to Readable Format

The secret message contains textual, numerical and special characters. So in this step, we convert the extracted and decompressed secret bits into a readable format. If we use the ASCII code of all the textual and numerical characters, then there is no need to send any additional information to the receiver for bit-to-character conversion, because ASCII codes are universally accepted. Sometimes if we have only a few characters in our secret messages, then using ASCII code is not advisable. Then, the sender creates his own encoding scheme. In this case, receiver must have the same encoding scheme so that he can further decode the secret messages from the decompressed bits.

Points worth remembering:
If ASCII code is not used for encoding then, the receiver must be aware of the encoding mechanism for characters to bits or vice versa.

12.4 PRACTICAL APPROACH TOWARDS STEGANALYSIS

As we saw in the previous chapter the steganalysis is nothing but the approach by which one can analyse the existence of the embedded secret message into the stego image. A steganography approach must be secure and robust enough against any steganalysis.

12.4.1 Cachin's Definition of Steganography

One can ensure the robustness against steganalysis according to Cachin's definition of steganography. According to Cachin's definition, we calculate the Kullback-Leibler distance between the probability distributions of the cover image and stego image. To make the algorithm more resistant to steganalysis we need to make the Kullback-Leibler distance as small as possible. We can achieve this by minimizing the distortion between the cover and stego image.

As we know that any steganography scheme can be modelled by two functions, one is Emb for embedder and other one is Ext for extraction. The work of Emb is to provide a stego image s for any given cover image c and message m. Let the cover image be drawn from all the available space of the cover with the probability distribution P_c, similarly, the distribution of the stego image is denoted by P_s. Comparison of these two probability distributions can be done using the Kullback-Leibler distance which is defined as:

$$D(P_c||P_s) = \sum_c P_c(c) log \frac{P_c(c)}{P_s(c)} \tag{12.6}$$

This distance will always be non-negative and it will be zero if $P_c = P_s$, which means the stego image is almost identical to the cover image and if $D(P_c||P_s) \leq \epsilon$, then Cachin called that system $\epsilon-$ secure system. It is up to the sender or steganography method developer to determine how he reduce the $D(P_c||P_s)$ value as small as possible. Here we are suggesting one method just to give an idea for the possibility of making a stego and cover image almost identical.

To achieve a very secure steganography method, it is recommended that the total number of secret bits must be smaller than the total number of embedding bits of the cover image. This type of constraint is useful for achieving secure steganograpgy methods with smaller Kullback-Leibler distance, as we can see in the following example.

Let us consider three pixel intensities of the cover image which are Pi_1, Pi_2 and Pi_3. We need to embed two secret bits b_1 and b_2 into these intensities. Here we are assuming that we are using the spatial domain for embedding. So in this example we are taking the first LSB of each pixel intensity for embedding and will let it be denoted by $LSB(Pi)$. Now we have to monitor two equations as given below:

$$b_1 = LSB(Pi_1) \ XOR \ LSB(Pi_2) \tag{12.7}$$

and

$$b_2 = LSB(Pi_2) \ XOR \ LSB(Pi_3) \tag{12.8}$$

We have few situations related to the above equations. If the intensities of the cover work satisfy both conditions, then there is no need to change the LSBs of the pixel intensities. If the first condition is true but the second one is false, then just flip the LSB of Pi_3. Similarly if the second condition is true but not the first one, then just flip the LSB of Pi_1. If both conditions are not true, then just flip the LSB of Pi_2. Here we can see that in the worst case, we also just change only one bit out of three. So in this way we can minimize the Kullback-Leibler distance and hence the opportunities for steganalysis.

Points worth remembering:
Cachin's definition inspires us to create more robust steganography algorithms to resist steganalysis.

Points worth remembering:
The Kullback-Leibler distance between the stego and cover image must be as small as possible.

12.4.2   LSB-Based Steganalysis

Here we are going to discuss a very basic approach for LSB-based steganalysis. As we know that this type of steganalysis comes under the category of targeted steganalysis because this approach is only focused on LSB-based steganography. Due to LSB embedding the histogram of pixel values shows some artifact characteristics. Whenever we embed the secret bit on a single LSB of each pixel, then even pixel values are either unmodified or increased by 1, whereas odd pixel values are either

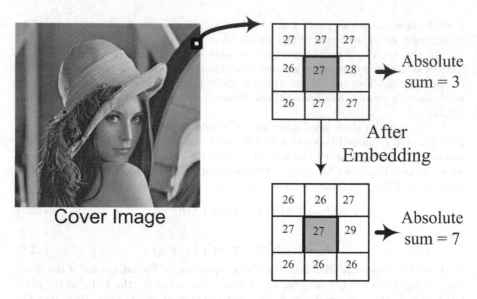

FIGURE 12.10 Example of LSB-based steganalysis.

unmodified or decreased by 1. So we can make a pair of values for gray scale pixels like $(2i, 2i + 1)$. This means for any value of i, the pixel intensity can only vary in $(2i, 2i + 1)$. This type of strategy can be used to create an estimator for steganalysis.

As we know that changes in LSBs are not symmetric. Flipping the LSB values of intensities is just like adding noise into a given cover image. That is why unknowingly we increase the average difference between the neighbouring pixel values. An attacker (for steganalysis) can just calculate the absolute sum of the difference between all pairs of neighbouring pixels of the cover image. The larger value of the absolute sum can indicate the existence of any hidden information.

One can understand this method by Figure 12.10. Here we can see that a region of average intensities in the cover image is chosen for the embedding of secret bits. Before embedding, the absolute sum of the differences between the center pixel (in this example 27) and its neighbour is 3. Once we embed the secret bits, then the same region will have an absolute sum of 7 which is comparatively high and pays attention to steganalysis. The effect of this change is negligible if we see this region visually before and after embedding.

SUMMARY

- There are three steps in embedding: *preprocessing, domain selection and bits embedding.*

- There are three sub-steps in preprocessing which are *lossless compression, ASCII code conversion and binary conversion.*

- *Lossless compression* of the secret message is divided into two types: fixed length encoding and variable length encoding.

- In *fixed length coding*, all the characters of the secret message are encoded using the same length of bit stream.

- In *variable length coding*, all the characters of secret the message are encoded using different lengths of bit streams.

- User-defined text to number mapping is more desirable from a security as well as embedding capacity point of view.

- *Shuffling* of secret bit streams is desired to provide confidentiality even after extraction of the secret bit stream.

- Embedding in all pixels is less secure but allows good embedding capacity.

- *Region-of-interest-based* embedding is appropriate when we have less but very important secret information to be embedded.

- Embedding in the frequency domain is robust enough because of the reversibility property of the transformation matrix.

- Extraction of secret bits from the spatial domain is much easier than the extraction from the frequency domain.

- If ASCII code is not used for encoding, then the receiver must be aware of the encoding mechanism of characters to bits or vice versa.

- *Cachin's definition* inspires us to create more robust steganography algorithm to resist steganalysis.

- The *Kullback-Leibler distance* between the stego and cover image must be as small as possible.

IV

Hybrid Approaches and Advanced Research Topics

Hybrid Approaches for Image-Based Security Using VC, Watermarking & Steganography

CONTENTS

S O far, we have seen both the theoretical and practical aspects of the three dimensions of image-based security techniques: visual cryptography, digital image watermarking and image steganography in detail. One can create a very efficient and secure method to protect images by choosing various combinations of these three techniques. In this chapter we will see different combinations of the approaches that we studied earlier. We will also discuss the basic development model of such hybrid methods. Readers can develop their own methods after understanding the basics in this chapter. The **Chapter Learning Outcomes(CLO)** of this chapter are given below. After reading this chapter, readers will be able to:

1. Understand the need for an hybrid approach.

2. Understand the basic development of a hybrid approach.

3. Develop his/her own efficient algorithm using different combinations of image-based security techniques.

4. Develop the MATLAB code for the implementation of the approaches.

13.1   INTRODUCTION

All the methods studied previously were developed to play a very fixed (specific) role in securing the images. Development of a visual cryptography approach is for

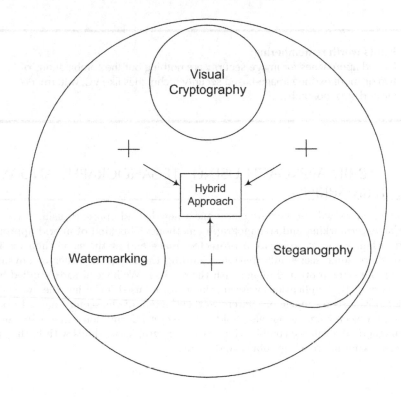

FIGURE 13.1 Relation of hybrid approach with other image-based security techniques.

sharing digital images securely. Not only the digital images but also the physical images can also be shared secretly using visual cryptography and can be recovered at the destination without any computation. Digital image watermarking is used to protect the ownership assertion as well as the integrity of the secret image. One can also recover the tampered portion of the secret image using watermarking. Steganography is used to protect the secret information behind the images. We can see that all approaches have their own fixed role, but what will happen if we combine the strengths of two or more image-based security techniques together? In this chapter we will see the various combinations of two or more image-based security approaches and their benefits in real life applications. In Figure 13.1 we see that the hybrid approach is a superset of other image-based security techniques like VC, watermarking and steganography. In upcoming sections, we will see how we can combine two or more techniques in order to develop a more secure and efficient image-based security approach.

Points worth remembering:
Hybrid approaches for image security are nothing but the combinations of two or more existing image-based security techniques like VC, watermarking and steganography.

13.2 HYBRID APPROACH USING STEGANOGRAPHY AND WATERMARKING

In this section we will see how we can make a new hybrid approach using a combination of watermarking and steganography methods. This kind of hybrid approach is useful when we want to create a protected image and at the same time we also want to share some secret information regarding the image or owner or anything else using the same protected image with the receiver. We have already studied how on the basis of the application, watermarking can be used for either the ownership assertion of the secret image or content authentication of the secret image. In both the cases the secret image can also hold some secret information using a steganography method. We will see combinations of steganography methods with both types of watermarking approaches: robust and fragile.

13.2.1 Combination of Robust Watermarking and Steganography Approach

Robust watermarking is used to ensure the ownership of the owner of the image. Ownership information may be any logo (binary or gray image) belonging to the owner. Sometimes the owner also wants to embed some additional textual information which may be related to the owner or his organization. In this scenario if we treat that textual information as an image, then embedding of that textual information will take a greater number of bits. To avoid such scenario we treat those textual messages as alphanumeric text only and apply a steganographic method to embed it. By this way we cannot only save the embedding space but also improve the imperceptibility. We can understand this scenario by Figure 13.2. Here we can see that we have textual information of three characters which is "ABC". If we convert it into the ACSII code and then convert it into eight-bit binary form, then the total number of required bits will be $8 \times 3 = 24$. At the same time if we create a document binary image for the same textual information and the size of the image is 32×64, the total number of bits required to embed this information is $32 \times 64 = 2048$. Now we can understand why embedding of textual information is better than embedding of document images.

FIGURE 13.2 Difference between text and document image.

Points worth remembering:
Embedding of a textual message is more space effective than a large document (binary) image.

If we want to combine the robust watermarking approach with steganography, then we need to select the proper image domain for both. We have already seen the significance of both the domains (frequency and spatial) in earlier chapters. Because a watermark must be robust enough against various kinds of intentional and unintentional attacks, it must be inserted into the frequency domain of the cover image. Because of block-based embedding into the frequency domain, we cannot insert more bits in this domain, hence we can choose the spatial domain for embedding of secret textual messages. This domain selection information can be seen in Figure 13.3. Here we can see that the frequency domain for the cover image is reserved for Robust watermarking whereas spatial domain for the cover image is reserved for steganography. In earlier chapters we saw many methods to embed a robust watermark into the frequency domain, so we can choose any one of the methods for embedding. Similarly embedding a secret message into the spatial domain is also not new for us. We have various approaches to do the same. Here unlike fragile watermarks we have textual information which can be compressed using any lossless compression mechanism to increase the embedding capacity of the cover image.

Points worth remembering:
We can reserve the frequency domain for watermarking and the spatial domain for embedding of secret message (steganography).

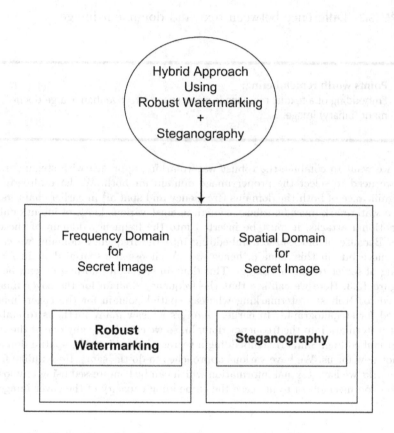

FIGURE 13.3   Domain selection for robust watermarking and steganography.

13.2.1.1 Steps to make a hybrid approach using robust watermarking and a steganography approach

Readers can develop their own way of embedding. Here we are describing some steps which may be followed to create a hybrid approach using robust watermarking and steganography.

1. We have three inputs available: watermark logo (binary image), textual information and cover image. First of all transform the cover image into its frequency domain. We can choose any reversible transformation methods such as discrete cosine transformation, discrete fourier transformation, etc. It is up to the reader's discretion to choose the transformation method.

2. Divide the obtained coefficient matrix into the non-overlapping blocks. Modify the coefficients according to the watermark logo bits and the selected algorithm.

3. After insertion of the watermark logo, take the inverse transformation.

4. Preprocess the textual secret information, e.g., compress it using any lossless compression method, take the ASCII value and finally convert it into the binary form.

5. Now we need to embed this preprocessed secret message into the LSBs of the pixel intensities of the cover image. As we know that the spatial domain is not robust against attacks so in order to protect the secret message we need to embed the same secret message in multiple positions of the cover image as shown in Figure 13.4. Here we can see that the secret message "ABC" is embedded into multiple locations of the spatial domain of the cover image. Suppose a binary bit stream of the secret message "ABC" is twenty four bits (eight bits for each character) long and if we assume that we are going to embed one bit per pixel, then we can divide the spatial domain of the cover image into non-overlapping blocks of size $m \times n$ such that $m \times n = 24$. Each block will contain the same secret message and if due to any reason some pixel of the cover image is altered during the transmission, then we can extract the secret message from other blocks. One can understand the benefits of the embedding of identical messages into multiple locations in Figure 13.5. Here we can see that the original Lena image is divided into four non-overlapping blocks and the same message "Hi.. I am Shivendra" is embedded into each block. Here three blocks are tampered with due to an intentional attack but one is safe. So we can extract the error-free message from the unaltered block. It will be more beneficial if we divide the image into a large number of blocks.

Points worth remembering:
A small secret message can be embedded into multiple locations of the spatial domain of the cover image.

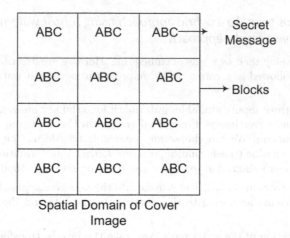

FIGURE 13.4 Embedding of secret message into multiple locations.

13.2.1.2 Effect of steganography on the watermarked image

Since we are applying steganography after the watermark embedding, our robust watermarking approach must be robust enough to tolerate the obvious alteration of the pixels done by the steganography approach. A wrong choice of steganography approach may decrease the imperceptibility of the watermarked image. So we need to pay attention to this also. We need to select a type of steganography approach that has very little effect on the watermarked image. Steganography has nothing to do with the security of the cover image; it is just an additional feature we are adding to watermarking by which we can also secretly share some textual information with the receiver.

Points worth remembering:
Steganography over watermarking can reduce the imperceptibility of the watermarked image.

13.2.2 Combination of a Fragile Watermarking and Steganography approach

A steganography approach can also be combined with the fragile watermarking approach. As we know that a fragile waermarking approach is used for content authentication so it is embedded into the spatial domain of the cover image. To not disturb the watermark bits, we can insert the secret message into the frequency domain of the cover image. We can see the domain selection in Figure 13.6. In this case we cannot embed multiple copies of the secret message due to space constraints

Hi..I am Shivendra   Hi..I am Shivendra

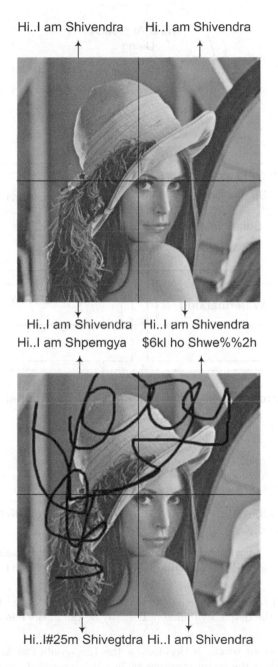

Hi..I am Shivendra   Hi..I am Shivendra
Hi..I am Shpemgya $6kl ho Shwe%%2h

Hi..I#25m Shivegtdra Hi..I am Shivendra

FIGURE 13.5 An example of embedding the secret message into multiple locations.

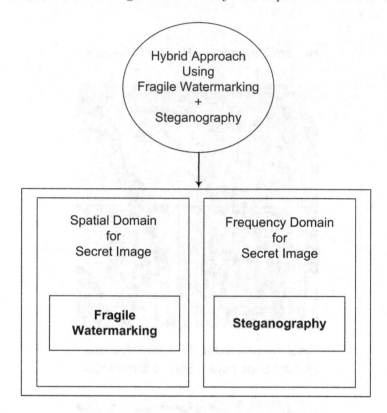

FIGURE 13.6 Domain selection for fragile watermarking and steganography.

in the frequency domain, but there is no need to embed multiple copies due to the robustness of the frequency domain.

13.2.2.1 Steps to make a hybrid approach using fragile watermarking and a steganography approach

In order to make a hybrid approach using fragile watermarking and steganography, one can follow the following steps. However, readers can also create their own efficient approach.

1. First of all take any reversible transformation of the image from the spatial domain to the frequency domain. Then divide the image into non-overlapping blocks.

2. The number of blocks is decided by the the number of bits of the secret message. Here the secret message again must have been preprocessed (compressed).

3. We embed all the secret bits into the frequency domain of cover the image using any suitable method.

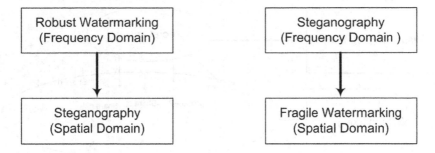

FIGURE 13.7 Valid sequences for steganography and watermarking.

4. Take the inverse transformation of the image. Now we need to embed the fragile watermark into the cover image in the spatial domain.

5. A fragile watermark may be generated by any self-embedding technique. Once we embed the fragile watermark, the protected image is ready.

13.2.3 Valid Sequence for Steganography and Watermarking

Till now we have seen how to apply both the techniques (watermarking and steganography) into the same cover image. Now the main question arises: Is there any valid sequence which must be followed for combining both techniques? The answer is "YES". Here we explain the proper justification for that. Suppose we are talking about the combination of robust watermarking and steganography. In this case if we apply steganography first into the spatial domain, then when embedding of the robust watermark into the frequency domain, the secret message may be altered. But in the reverse approach, if we embed the robust watermark first in frequency domain and then we embed secret message into the spatial domain, it will hardly affect the watermark. Because the watermark is robust, it can tolerate the changes done by the steganography. Now suppose we are taking a combination of fragile watermark and steganography. Here if we embed the fragile watermark first into the spatial domain and then embed the secret message into the frequency domain, then changes done during steganography will be reflected in the fragile watermark which should not have happened. That is why we embed the secret message first into the frequency domain of the cover image and then embed the fragile watermark into the spatial domain of the cover image. In Figure 13.7, we see the valid sequence for robust/fragile watermarking and steganography. We can also see in Figure 13.7 that all frequency domain computations are followed by spatial domain computations.

Points worth remembering:
Operations in the frequency domain must be followed by operations in the spatial domain.

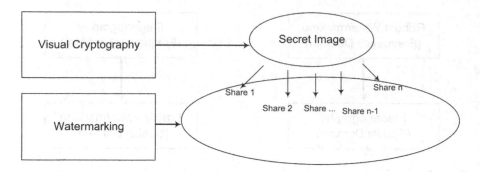

FIGURE 13.8 Hybrid approach using visual cryptography and water-marking.

13.3 HYBRID APPROACH USING VISUAL CRYPTOGRAPHY AND WATERMARKING

In this section we will study how many combinations of hybrid approaches can be made using visual cryptography and watermarking. Actually we can further divide visual cryptography approaches into two sub-categories on the basis of share appearance: VC with halftone shares and VC with multitone shares. We have already seen that watermarking can also be categorized into two types: fragile and robust watermarking. Now we will discuss which combination of VC and watermarking approaches is valid and suitable for making better image-based security techniques.

Actually watermarking is used for ownership assertion as well as content authentication while visual cryptography is used to share the secret image into two or more shares. The question arises of how to combine these approaches together. Here we suggest a possible combining method but readers can make their own. First of all we can apply a visual cryptography approach to the secret image to make two or more shares. If the secret image is halftone, then the generated shares will also be halftone and if secret image is multitone then generated shares may be halftone or multitone. After getting the shares, we need to protect them by using fragile watermarking or a robust watermarking approach. This process is shown in a Figure 13.8 where VC is applied to the secret image and watermarking is applied to the generated shares. In upcoming sections, we will see all other combinations of VC and watermarking approaches.

13.3.1 Combination of VC with Halftone Shares and a Watermarking Approach

By VC with halftone shares we mean VC approaches which generates binary shares. So in this section we will see how the watermarking approach is applied to shares to protect them. Applying robust watermarking on halftone (binary) shares is meaningless because in halftone shares, we have only two combinations of intensities either black or white. If we take the frequency transformation of the binary shares to embed the copyright information, the resulting shares will not remain halftoned. Now we have only one combination left which is VC with a fragile watermark.

Points worth remembering:
Halftone shares are not suitable for robust watermarking.

13.3.1.1   Protection of halftone shares using fragile watermarks

As we know that shares are very important because they carry secret information. So integrity verification of shares is very necessary at the receiver end. All received shares must be authentic and there should not be any modification during the transmission. We can achieve this by applying fragile watermarking on halftone shares. As we know that halftone shares have only one bit depth at each pixel location so we need to apply a block-based fragile watermarking approach.

Steps to embed a fragile watermark into the halftone shares:

1. First of all divide a halftone share into non-overlapping blocks of size $m \times n$.
2. Select the one random location in a block in which the generated watermark will be embedded.
3. Apply a any self-embedding algorithm to the remaining $m \times n - 1$ pixels of the block to generate single authentication bit.
4. Embed the generated authentication bit into the reserved location of the block.
5. Apply the same procedure for all blocks of the share.

One can understand the aforesaid algorithm using Figure 13.9. Here a halftone share is divided into blocks of size 3×3 and the authentication bit for each block is generated by eight bits of the block. This authentication bit is embedded into the remaining single bit position of the block.

Points worth remembering:
Halftone shares have a single bit per pixel, so block-based fragile watermarking is used in this case.

13.3.1.2   Effect of a fragile watermark on shares

The effect of fragile watermark embedding into the halftone shares directly depends on the size of the non-overlapping block. If we have a large size block, then it will have less effect on the share and vice versa. Since halftone shares are already noisy in nature, one cannot judge the change due to watermark embedding just by the human

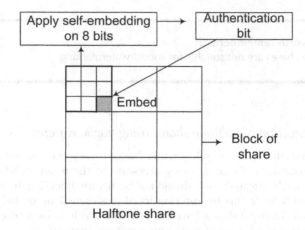

FIGURE 13.9 Embedding of fragile watermark into the halftone share.

visual system. Changes in shares may be reflected in the recovered secret when we superimpose all the halftone shares. Watermark embedding may reduce the contrast of the recovered secret. Hence readers must consider this point during developing of any algorithm. We should choose such a position for embedding the authentication bit in the block which cancels the effect when stacked with all remaining shares. For achieving this goal we may choose different position for embedding in the blocks of different shares. It will also increase the security.

Points worth remembering:
By embedding the fragile watermark into halftone shares, the contrast of the recovered secret may decrease.

13.3.2   Combination of VC with multitone shares and watermarking approach

A VC approach which provides multitone shares as output is known as VC with multitone shares. Here we assume that secret image definitely be also multitone in nature. We can apply any of watermarking approach(Robust or Fragile) on multitone shares. In this section we will see combination of both watermarking approaches with VC.

13.3.2.1   Protection of a multitone shares with a robust watermarking approach

As we know that multitone shares may be gray or colored intensities. Each intensity of a pixel may be, either of eight-bits or twenty-four bits. So, in this case, we can treat a share just like a cover image for watermark embedding. But this process is not as simple as it looks because shares are not standalone entities. They are dependent on each other. At the same time, we also have to pay attention to the quality and contrast of the recovered secret image using all multitone shares.

Steps to embed a robust watermark into the multitone shares:

1. In this case we cannot separate the process of visual cryptography and embedding of a robust watermark into the shares. As we have already studied in earlier chapters we are free to do computations in case of multitone secret and shares. So we will take the benefit of that. Initially we need to decide how many bits among eight will be dedicated for storing the information of share (i.e., secret) and how many bits will be dedicated for storing the copyright information.

2. Let us have n number of multitone shares and a secret of size $M \times N$. We are taking k number of bits from each pixels of a share to secretly share the secret image. So we have in total $k \times M \times N \times n$ number of bits to encode the information of secret image having $8 \times M \times N$ number of bits. We need to maintain the relation $8 \times M \times N << k \times M \times N \times n$. Since in multitone visual cryptography, we are allowed to do computation, one can easily make a visual cryptography algorithm with this constraint . As we have $8 - k$ number of bits left in the each pixel of the share, so total number of bits per share which may be used to embed the copyright information is $(8 - k) \times M \times N$.

3. We have to create another matrix of size $M \times N$ for each share in which a pixel will be of $8 - k$ bits. Now these matrices will be treated as separate images and we will apply our robust watermarking approach on it to hide the copyright information.

4. Once we embed the copyright information, just append the $8 - k$ bits of each pixel with the remaining k bits of the corresponding pixel. In this way we again get the eight bit pixel in which k bits contain the secret image information and $8 - k$ bits contain the copyright information.

One can understand this whole procedure of having a combination of robust watermarking and visual cryptography through Figure 13.10. Here the procedure is shown with only a single pixel of secret, shares and copyright logos. One can see that eight bits of a pixel of the shares are divided into two parts. One part is responsible for encoding the secret image whereas the other part is responsible for embedding copyright logo. The interesting thing to understand here is eight bits of a single pixel of a secret image are encoded with k bits of the corresponding pixel of all shares whereas $8 - k$ bits of a pixel of a share are responsible for embedding the copyright logo for the same share only.

We have shown a very small example in Figure 13.11 of embedding by robust watermarking into the multitone shares. Actually this example is based on the algorithm shown in Figure 13.10. Here we are taking a secret image of size 2×2. We have three multitone shares for this secret image. We have split the eight bit binary stream of each share into two parts. One part (six bits) of each pixel will be used to encode the secret image whereas the image generated by the remaining bits (two bits) of each pixel of the shares will be used to embed the copyright logo for an individual share. We transform the image generated by two bits into the frequency

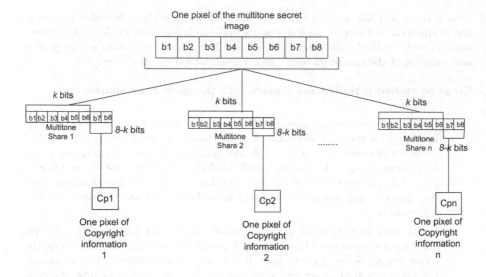

FIGURE 13.10 Embedding of robust watermark into the multitone share.

domain before embedding the copyright logo so that we can maintain the robustness.

Points worth remembering:
A single pixel of the share will contain both types of information: secret image and copyright logo.

13.3.2.2 Protection of multitone shares with a fragile watermarking approach

Ownership of shares can be verified by robust watermarking but what about content authentication? Shares may be tampered with during transmission so they must be protected by a fragile watermarking scheme. In the previous section, we saw the protection of halftone shares using block-based fragile watermarking. In that case, we did not have any other option except block-based embedding because in pixels of one bit (binary) we cannot embed any other information. Here we have multitone shares in which a pixel may be either eight bits (for gray scale) or twenty-four (for color) bits. Also, we saw that embedding of robust watermarking cannot be entirely isolated from the share generation process using visual cryptography. Similarly embedding of fragile watermarking can also not be isolated from the share generation process. We have to pay attention to the embedding locations in each pixel for fragile watermark bit embedding during the share generation process. The following are the steps for doing the same. This is not the only process, and readers can create their own effective approach.

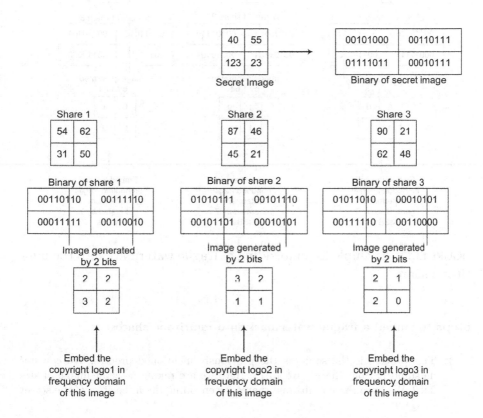

FIGURE 13.11 Example for embedding of robust watermark into the multitone share.

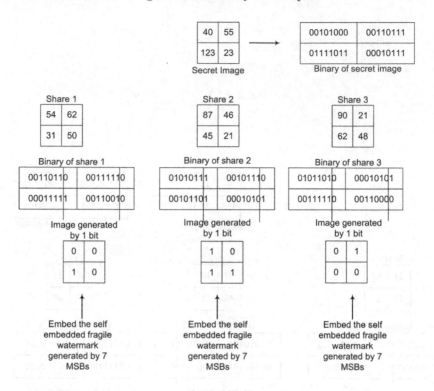

FIGURE 13.12   Example for embedding of fragile watermark into the multitone share.

Steps to embed a fragile watermark into multitone shares:

1. This process is the same as the previously mentioned process. We can understand it from the Figure 13.12. Here we are taking a secret image of size 2×2. We have three multitone shares for encoding the intensities of the secret image.

2. We split the eight bits of each intensity of a share into two parts. The first part is seven bits which will be used to encode the secret image whereas the second part which is a single bit will be used to embed the authentication bit for that pixel only. This is just an example, and there is no hard and fast rule to take a single bit for authentication. We can choose more than one bit also for better integrity verification.

3. Now we can apply any self-embedding approach on the seven bits of all pixels in order to generate the authentication bit.

4. This generated authentication bit will be placed according to the dedicated LSBs of the corresponding pixel.

Using a pixel-based fragile watermarking approach, we can achieve good authentication or tamper detection results but they may reduce the quality of the recovered secret (100% recovery of secret is not guaranteed). Generally, in multitone VC, computation is allowed hence we can regenerate the secret image with full accuracy anyway. Block-based fragile watermarking is also suitable for multitone VC. Here we can achieve good result with less complexity.

A one point we may think that can we make a fragile watermark with recovery capabilities for the shares. We can almost recover the visual content of the meaningful multitone share using a fragile watermark but what about the content of the secret image which was hidden behind the shares? For decoding the secret image using shares at the receiver end, we need to extract encoded bits in shares, otherwise we would not be able to recover the secret image.

Points worth remembering:
A fragile watermark with recovery capabilities is not suitable for multitone shares.

13.4 HYBRID APPROACH USING VISUAL CRYPTOGRAPHY AND STEGANOGRAPHY

A hybrid approach using visual cryptography and steganography is just like the hybrid approach using visual cryptography and fragile watermarking approach. There may also be two other combinations: VC with halftone shares and steganography and VC with meanngful shares and steganography. As with fragile watermarking, a secret message in steganography can also be embedded into the spatial domain of the halftone or multitone shares. We also need to make sure that embedding does not degrade the quality and contrast of the recovered secret shares. We can follow exactly the same steps mentioned above for a hybrid approach using VC and a fragile watermark. There will be only one basic difference. In the case of fragile watermarking, the authentication bit (watermark) is generated using a self-embedding technique so there is a strong dependent relationship between the MSBs of the share and authentication bits. Whereas in the case of steganography, the secret message and share are both entirely independent of each other. We simply preprocess the secret message and embed it into the shares. If the shares are halftone in nature, then we use block-based embedding of the secret message otherwise pixel-based embedding.

Points worth remembering:
A hybrid approach using VC and steganography is just like a hybrid approach using VC and fragile watermarking.

13.5 HYBRID APPROACH USING VISUAL CRYPTOGRAPHY, WATERMARKING AND STEGANOGRAPHY

This is the time to combine all the image-based security approaches (Visual cryptography, watermarking and steganography) together. There may be many combinations of the techniques for example, VC may be either with halftone shares or with multitone shares, a watermark may be either fragile or robust or they also may be either block-based or pixel-based, so one can pick any of the combinations and proceed further. Here we mention one method to show how it is possible to combine all in one. Readers can create their own effective algorithm.

13.5.1 Combination of Halftone VC with Fragile Watermarks and Steganography

As we already know that embedding of any information into shares must not affect the quality and contrast of the recovered secret while stacking all the shares. So we need to create an algorithm that is capable of doing the same. We have seen that a halftone share contains pixels of a single bit, so in order to embed the information we need to do block-based processing. In this section we are going to embed a secret message and watermark in all the shares, hence we need to preprocess the halftone shares accordingly. Here we provide the steps to create such an algorithm.

13.5.1.1 Steps to embed a fragile watermark and steganography into the halftone share

1. Here we are assuming that all meaningful halftone shares and the secret image are preprocessed as suggested in the earlier chapters of visual cryptography. A preprocessed image has blocks of size 2×2 containing either all black or all white pixels.

2. Now we refer to Figure 13.13, where we can see that the secret image and all halftone shares are divided into non-overlapping blocks of size 8×8. This example shows processing of only one block.

3. First of all we randomly choose any black block of size 2×2 from the secret image in the bigger block of size 8×8. Now we map the same location in all shares as we assume that all shares and the secret are of the same size.

4. In this example we take four shares. During embedding of the secret messages and watermarks in the shares, we have to ensure the secret image quality. That is why here we are embedding an auxiliary black pixel in each block of the share.

5. We can see in Figure 13.13 that with the auxiliary black pixel we are also embedding watermark Wc, secret message Sec and authentication bit Au in a fixed manner. This sequence relates for all the successive shares.

6. As we know that except for the auxiliary black pixel, all other information (watermark, secret message etc) may be either black or white. Hence due to rotation of the block in each share, the auxiliary black pixel will dominate and hence when we stack all the shares, we will finally get an all black block of size 2×2.

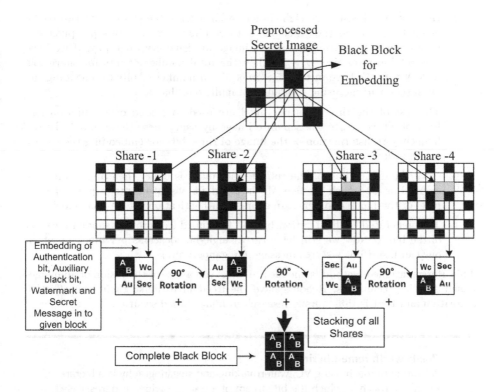

FIGURE 13.13 Example of embedding of fragile watermark and a secret message into halftone share.

13.5.2 Combination of Multitone VC with Fragile Watermarks, Robust Watermarks and Steganography

Unlike halftone shares, in multitone shares, we have the freedom to embed many things simultaneously because here each pixel of any share is of either eight or twenty four bits. In Figure 13.14 we can see one method to embed the fragile, robust watermark and secret message (steganography) simultaneously into the multitone share. Here are the steps to do this.

1. Here we are dividing the eight bits of each pixel of the shares into two parts, one is five bits and the other one is three bits. These five bits per pixel will be dedicated to sharing the secret image in visual cryptography. If we have a secret image of size $M \times N$, then the total number of bits for sharing is $M \times N \times 8$. We have in total $5 \times M \times N \times n$ number of bits for encoding the given secret image using n number of multitone shares.

2. The rest of the three bits per pixel are used to embed other information. First of all we create the separate image by using these three bits. Take the frequency transformation of the image of three bits for embedding the robust watermark.

3. Once the embedding of the robust watermark is done, then apply any self-embedding approach on five MSBs (used for visual cryptography) of each pixel of the block and generate a single authentication bit for that block.

4. This generated authentication bit can be embedded into the shaded location of the block. Now we still have three single-bit locations per block in which one can insert one's secret message (steganography).

This aforementioned method is one idea to make you aware of the fusion of all image-based security techniques. Readers have the freedom to make their own effective algorithm and that is thus pursue research in image-based security.

Points worth remembering:
A hybrid approach using VC, watermarking and steganography is a hierarchical approach in which the bit stream of a pixel is divided progressively to embed the secret, watermark and secret message, respectively.

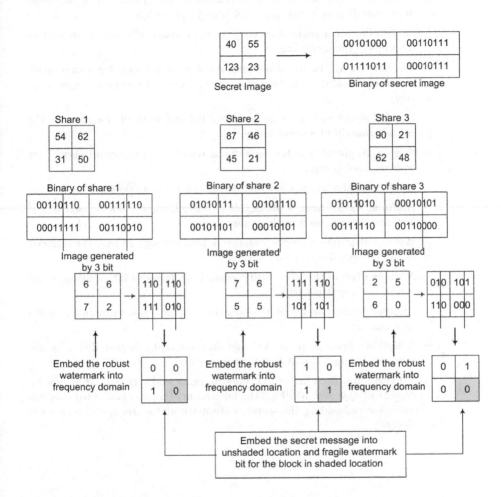

FIGURE 13.14 Example for the embedding of fragile, robust watermark and secret message into multitone share.

SUMMARY

- *Hybrid approaches* for image security are nothing but the combinations of two or more existing image-based security techniques such as VC, watermarking and steganography.

- There are three sub-steps in preprocessing which are *lossless compression, ASCII code conversion and binary conversion*.

- Embedding of a textual message is more space effective than a large document (binary) image.

- We can reserve the frequency domain of a cover image for a watermark and the spatial domain for embedding of a secret message (steganography).

- A small secret message can be embedded into multiple locations in the spatial domain of a cover image.

- Steganography over watermarking can reduce the imperceptibility of the watermarked image.

- Halftone shares are not suitable for robust watermarking.

- Halftone shares have a single bit per pixel, hence block-based fragile watermarking is used in this case.

- When embedding a fragile watermark into halftone shares, the contrast of the recovered secret may decrease.

- A single pixel of the share will contain both types of information: *secret image and copyright logo*.

- *A fragile watermark with recovery capabilities* is not suitable for multi-tone shares.

- A hybrid approach using VC and steganography is just like a hybrid approach using VC and fragile watermarking.

- A hybrid approach using VC, watermarking and steganography is a *hierarchical approach* in which the bit stream of a pixel is divided progressively for embedding the secret, watermark and secret message, respectively.

CHAPTER 14

Protection of Other Multimedia Objects Using Image-Based Security Techniques

CONTENTS

N OW at this point we can assume that readers have sufficient practical and theoretical knowledge about the three types of image-based security techniques: visual cryptography, digital image watermarking and steganography. These security approaches are not only able to secure the images but they can also be used to secure other multimedia objects like speech, videos, 3D images, etc. In this chapter we will see what other modifications are required in current image-based security techniques in order to secure other multimedia objects. These issues can be treated as areas of advanced research. After reading this chapter, readers will be familiar with various related fields and can protect all types of digital objects. The **Chapter Learning Outcomes(CLO)** of this chapter are given below. After reading this chapter, readers will be able to:

1. Develop his/her own audio watermarking, audio secret sharing and audio-based steganography approaches.

2. Develop his/her own video watermarking, video secret sharing and video-based steganography approaches.

3. Understand a few other advanced research areas related to multimedia security.

14.1 INTRODUCTION

Till now we covered three security techniques: visual cryptography, watermarking, steganography which are basically used for images. What do you think about using the same techniques for protecting other types of objects? This can be quite interesting and useful also. We will see security for all multimedia objects one by one as well as some introductory material for other advanced research related to the field of multimedia security. Readers can see the various research techniques for protecting multimedia objects based on image-based security techniques in Figure 14.1. This is not the complete list, one can explore more about the multimedia domains. If our concept related to image-based security techniques is clear, we can create our own algorithm to protect any multimedia object present in the real world. A very basic multimedia object is audio/speech. When we deal with its offline version, then it is called audio otherwise its online or runtime version is called speech. One can import the concepts of watermarking, secret sharing and steganography methods for security of this type of multimedia object. Similarly video is another important multimedia object, which must be protected sometimes. One can also import the concepts of image-based security techniques to protect it. Apart from these research domains, we can explore many other domains like 3D (dimensional) image protection. Till now whatever techniques we have seen were all related to 2D image protection only, but we can expand our algorithm to protect 3D images also. Image-based security techniques can also be used to protect telemedicinal data and biometric data. We can explore many other fields in which image-based security techniques can be used. We will read a brief introduction related to all the aforementioned areas in the forthcoming sections.

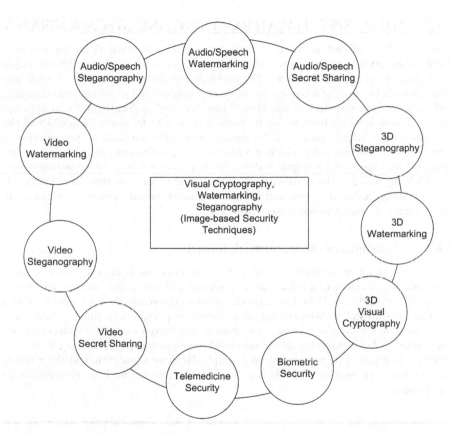

FIGURE 14.1 Various research domains for protection of multimedia objects based on image-based security techniques.

Points worth remembering:
Image-based security techniques can also be used to protect other multimedia objects like audio, video, etc.

14.2 AUDIO/SPEECH WATERMARKING AND STEGANOGRAPHY

Based on the application, we have seen two types of watermarking for the two-dimensional images: robust watermarking and fragile watermarking. We can apply these watermarking techniques for the protection of audio/speech also. Fragile watermarks will be used for integrity verification of audio whereas robust watermarking will be used for ownership verification. These types of applications are in high demand in broadcasting frequencies in radio FM or AM to prove ownership of the broadcasting authority on the audio signal. Sometimes we need to protect speech which may be used in court evidence without any modifications. In this case, speech must be protected with a fragile watermark. One can also make a recording device in which our watermarking algorithm is hard coded and at the time of recording, it automatically inserts a robust and fragile watermark into it. There are many more applications of audio/speech watermarking.

14.2.1 Selection of the watermark format

In the case of image as secret, because we have only one option for insertion i.e., copyright logo in cases of a robust watermark and self-embedding watermark in case of a fragile watermark. If we have speech/audio as the secret then we may have two options for embedding. We can embed a separate copyright logo (image) or a copyright audio in the case of a robust watermark. Similarly in a fragile watermark, we can embed either a self-embedded watermark or separate speech/audio for authentication. In Figure 14.2 we can see an example. Here we assume that at the receiver end, we have a processing unit which will perform extraction for both formats (audio and image).

Points worth remembering:
In case of Audio watermarking we have freedom to choose the watermark format(either audio or image).

14.2.2 Audio Robust Watermarking

Before proceeding to audio watermarking, we must know the format of the audio or speech file. Actually unlike images which are two-dimensional matrices, speech/audio file is one-dimensional array. As in an image matrix, where its element represents

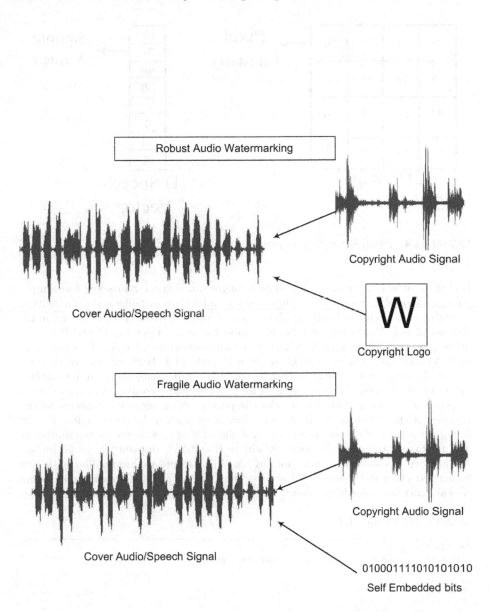

FIGURE 14.2 Freedom to choose watermark types (image or audio).

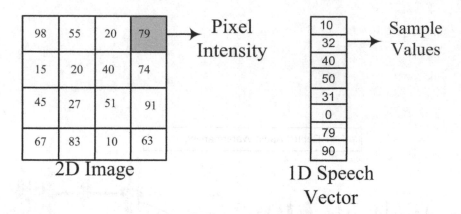

FIGURE 14.3 Pixel intensity versus speech sample.

the pixel intensity, similarly, in a speech vector, an element represents the sample value as shown in Figure 14.3. These sample values are just like audio intensities. A high sample value indicates a high pitch of audio whereas a 0 sample value indicates silence. Whenever we take any audio file as an input to MATLAB or any other processing software, it is converted into one-dimensional array. A sample code in MATLAB for recording speech is given in code 14.1. Here we can see that we decide the frequency of the recording by variable Fs. It means there are Fs number of samples per second. This example is for a five-second recording. The recording is stored in a variable y. Now y is the most important thing because it contains all the samples of speech in the form of a one-dimensional array. In this example we are fixing the format of speech as an unsigned eight bit integer. It means the minimum and maximum values of the samples will be 0 and 255, respectively, as an image. We are free to choose any other format also but because we are familiar with this format, we chose it. Once we get the cover signal vector and watermark signal vector we can start our embedding algorithm.

MATLAB Code 14.2

```
%MATLAB Code 14.1
% This MATLAB code is for recording the speech signal
clc
clear all
close all
 Fs = 11025;
  y = wavrecord(5*Fs, Fs, 'uint8');
   wavplay(y, Fs);
```

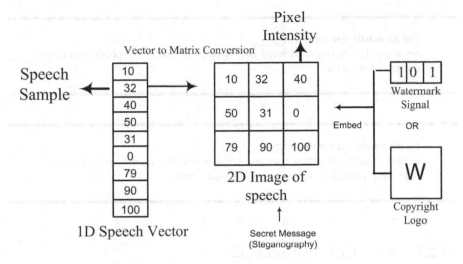

FIGURE 14.4 Process for robust audio watermarking.

14.2.2.1 *Steps to embed robust audio watermarks or steganographic messages*

Let the lengths of the cover speech vector and watermark vectors be $|y|$ and $|w|$, respectively. Our assumption is $|y| >> |w|$. The following are the steps for embedding the robust watermark into the cover speech vector:

1. First of all make the two-dimensional matrix of the vector y either row wise or column wise.

2. Here we assume that either we are taking binary image as copyright logo or a very small speech signal as copyright information. This small speech is to be converted into its binary version.

3. Now we have the same situation as in image watermarking. The cover speech is treated as cover image and the watermark signal is treated as a binary logo. Now just apply any previously discussed robust watermarking algorithm for the images on it. We can better understand this scenario by Figure 14.4. Here we see that an audio vector is converted into a two-dimensional matrix. We have two options for embedding, either a binary copyright logo or a binary speech signal. Next, we will take any frequency transformation of the cover matrix and embed the watermark signal. In Figure 14.4 we also see that we have the ability to embed any secret message as steganography in the frequency domain of the signal. These are just like the previously described approaches, but the basic difference here is that the cover is now speech not an image.

Points worth remembering:
An image is a two-dimensional matrix whereas audio or speech is a one-dimensional vector.

Points worth remembering:
Readers can choose their own recording format for speech and any copyright logo (image) or speech signal as watermark.

14.2.3 Audio Fragile Watermarking

We can also embed the fragile watermark for content authentication of speech or audio. The process is just like the image-based fragile watermark. Here we embed the watermark into the LSBs of the samples. Because readers are familiar with the format for speech (here we are taking uint8), they can easily convert the samples into binary format. We can follow the below steps to embed the fragile watermark or a steganography message into the cover speech/audio.

14.2.3.1 Steps for embedding a fragile watermarks or steganography messages

1. We first apply the process to convert the audio vector into a two-dimensional matrix.

2. Since here we are taking unsigned eight bit integer format, the range of samples lies from 0 to 255. Now we dedicate the last LSB of each sample for watermark/steganography message embedding.

3. To generate the fragile watermark, apply any self-embedding algorithm on the other seven bits per sample. This self-embedding approach must generate a single bit output.

4. This generated single bit will be embedded into the first LSB of each sample.

5. One can understand the whole procedure by Figure 14.5. Here we can see that a single authentication bit is generated for each sample using a self-embedding approach. These generated authentication bits are now embedded into the first LSBs of the samples. We can also embed any secret message in this place if we are dealing with steganography.

6. In order to prevent degradation of the cover signal because of embedding, we can think about block-based embedding of the fragile watermark.

Here we have shown just a very basic, sample approach for audio watermarking or steganography. With these basics, readers can create their own efficient approaches.

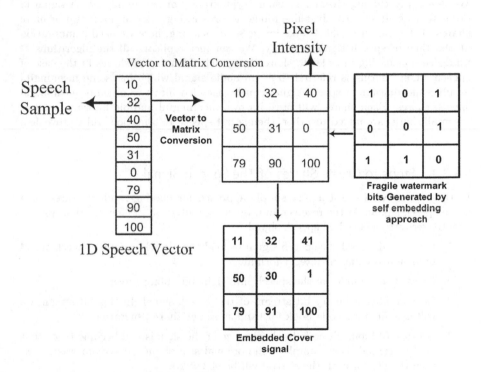

FIGURE 14.5 Process for fragile audio watermarking.

Points worth remembering:
Fragile watermark bits are generated using any self-embedding approach applied on bits of audio samples and embedded into the LSBs of the samples.

14.3 AUDIO/SPEECH SECRET SHARING

We can apply the algorithm for visual cryptography on an audio/speech signal in order to share it secretly. Just like image secret sharing. we can make two or more shares of the speech signal. Unlike image secret sharing, here we need computation at the time of speech signal recovery. We can just replicate all the algorithms of image secret sharing in order to share the speech. Meaningful shares in the case of speech secret sharing is referred to as an audio signal which has some meaningful audible information. A speech signal can be encoded into two or more meaningful speech shares. Each shares will have its own meaningful audible information, but when all shares are mixed together, then secret speech information will be revealed.

14.3.1 Steps to Create Shares of The Speech Signal

In this section we suggest a very simple approach for making random shares for a given speech signal. At the receiver end, by doing a little bit of computation, we can exactly recover the original speech signal.

1. First of all record the speech signal or load the already existing speech signal in unsigned eight bit integer format.

2. Convert all samples of the speech into eight bits binary form.

3. Now we have a binary bit stream of the speech signal. In this bit stream we will use Shamir's code book for random secret share generation.

4. Because of Shamir's code book, the size of the signals will become twice that of the original. For example if the original size of the bit stream was $1 \times n$, then the size of both the shares will be of $1 \times 2n$.

5. At the receiver end, we just do EXOR of both the shares. In the resulting bit stream, just replace all pairs of 0s by a single 0 and all pairs of 1s by a single 1. By this way we will get the exact secret speech.

One can understand the aforementioned method by Figure 14.6. We can see that first of all, the samples of the speech vector are converted into binary form. After that we apply Shamir's basis matrix on this binary bit stream to generate two random shares. We can again convert the binary shares into decimal form so that they make some random sound. At the receiver end, when we EXOR both shares, we get various pairs of 1s and 0s. At the end, we just replace all the pairs of 1s and 0s by, single 1 and 0 bit, respectively. In order to create meaningful shares, we need some meaningful signals which could be made shares. After that, the same procedure of random share generation will be carried out. Once we get the random

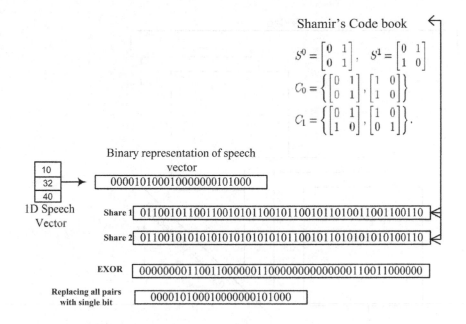

FIGURE 14.6 Audio secret sharing with two random shares.

shares, just hide the content behind the meaningful signals (shares) in such a way that after embedding we still hear the actual meaningful content of the shares.

Points worth remembering:
Shamir's basis matrix can be applied on the binary stream of speech for developing a basic audio secret sharing approach.

14.4   VIDEO WATERMARKING, STEGANOGRAPHY AND SECRET SHARING

Just like images and audio, we can also apply all the security techniques on video. Before proceeding to the security techniques for video, we must recall the basics of video. Video is nothing but a sequence of images which are also called frames. Here we first mention some basic terminology related to video.

14.4.1   Basic Terminology for Video

Frame size: Frame size is the size of the image shown in a frame. Generally it is denoted by $W \times H$.

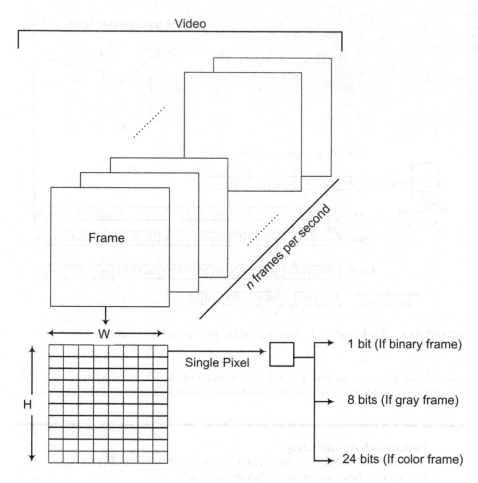

FIGURE 14.7 Fundamentals of video and its components.

Pixels per frame: Total number of pixels present in the frame. Let pixel per frame be N then $W \times H = N$.

Color depth: Color depth denotes the number of bits required to represent one pixel of the frame. If a frame is binary, gray or color, then the color depth will be one, eight and twenty-four, respectively. It is denoted by CD.

Bits per frame: Total number of bits required to represent a single frame. It can be calculated by $W \times H \times CD$ or $N \times CD$.

Frame rate: It is a very important parameter for video. It shows the number of frames which to be displayed in one second. It is denoted by F_r.

One can understand the above mentioned properties and components of video by Figure 14.7. Here we have broken the video into frames then pixels and then bit levels.

Example 14.1: Suppose we have a video of T seconds, then how many bits will be required to show complete video?

If the frame rate of a video is F_r then the total number of frames in T seconds is $F_r \times T$. Let the size and color depth of a frame be $W \times H$ and CD, respectively, then the total number of required bits is $F_r \times T \times W \times H \times CD$.

This example is just to review the basics of video and make you all comfortable enough to play with pixels and bits of the video.

14.4.2 Process for Video Steganography/Watermarking

As we have seen, video is just a sequence of images. It means if we process a video in a frame by frame manner, then we can simply apply image-based security techniques on frames and finally combine all the frames to get video. We can understand this process by Figure 14.8. Here first of all, all the frames of the video are separated and then we apply any image-based security technique on each frame. Finally we combine all the frames in order to get protected video.

Points worth remembering:
Image-based security techniques are applied on each frame of video and finally all frames are combined together in order to achieve video steganography/watermarking/secret sharing.

14.4.3 Enhancing the Security of Video Watermarking/Steganography

We have seen a very straight forward method to make an algorithm for video watermarking/steganography or secret sharing. This is just for a basic understanding so that readers pursue different direction to create new algorithms. Here we are suggesting a new dimension to enhance the security of video watermarking or steganography. By this way readers can think independently to develop efficient and secure methods. If we do not embed the watermark or secret message for steganography in all the frames, but we choose the frames of video, randomly for embedding, then we can enhance the security. It is a very trivial improvement in the basic algorithm but it can enhance the security considerably. Only the sender and receiver will know about the embedded frames. By this way we can improve the imperceptibility also because only a few frames will be altered due to additional message embedding.

In the MATLAB code 14.2, we can see a code for reading a video file. In this code we divide the video into frames, and thereafter the frames are modified. After modification, all frames are combined together. This is just a skeleton code. Readers can modify it as per their requirements.

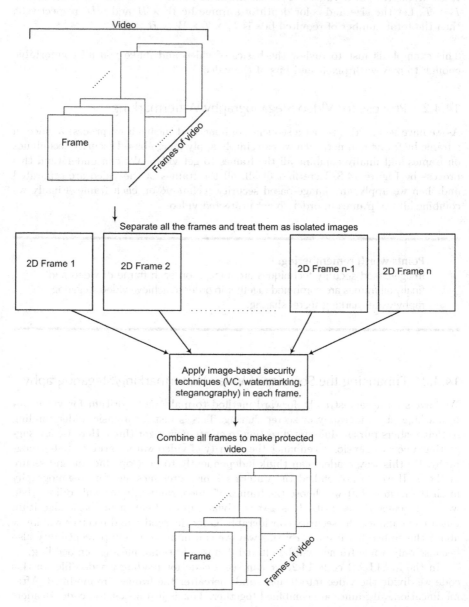

FIGURE 14.8 Applying image-based security techniques on video.

Points worth remembering:
Randomly selection of frames in video for watermark or secret message embedding can enhance the security.

MATLAB Code 14.2

```
%MATLAB Code 14.2
% This MATLAB code is for reading a video file, dividing it into frames, modifying the
    frames and again combining it.
clc
clear all
close all
a=mmreader('Video.avi');// Read the video file
b= a.numberofframe;
c=read(a);
mov(1:b)=struct('cdata',zeros(a.height,a.width,3,'uint8'),'colormap',[ ]);
for k= 1:b
    mov(k).cdata= c(:,:,:,k); // Save its all frames.
end
mov2=mov;
for k=1:b
    w= frame2im(mov(k));
    s=255-w; // This is the frame modification portion. Here we are taking the negative
        of frame. Embedding code will be inserted here.
    mov2(k)= im2frame(s);
end
imshow(mov2(k).cdata)
movie2avi(mov2,'alte.avi')// Frame to video conversion.
```

14.5   3D WATERMARKING

So far we covered watermarking for 2D images, audio and video in the book but there is still something missing i.e. 3D watermarking. Although this topic is outside of the scope of this book, we can still provide some basic introductory knowledge on this domain.

14.5.1   Introduction

Nowadays there are many applications involving 3D models. Due to the increasing power of graphics even in personal computers, 3D models have become popular multimedia objects now. These objects are increasingly used in gaming and animation applications. Hence it is necessary to have an intellectual property right protection mechanism for these type of models. This is the latest area of research providing a 3D watermarking algorithm for ownership verification and content authentication. We have already seen many methods to embed a watermark into the 2D image, audio and videos, but the 3D watermarking problem is more complex and difficult than others, so little research is available in the literature.

2D images are nothing but bi-dimensional regularly sampled collections of intensities whereas 3D objects are entirely different from 2D. We cannot say that 3D images are just an extension of 2D because in this manner video can be multi-

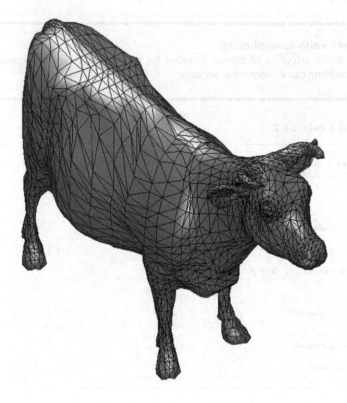

FIGURE 14.9 Example of 3D model with triangular meshes *(Courtesy of MPII by the AIM@SHAPE-VISIONAIR Shape Repository).*

dimensional representation of 2D images, but video is not a 3D model. 3D models can be represented by 3D polygonal mesh or set of implicit surfaces. Polygonal meshes are supposed to be triangular meshes because a triangle can be considered as the smallest polygon in geometry. One can see Figure 14.9 where a 3D object is shown. Here we can see that this object is made up of many vertices and these vertices are used to create triangular polygons. Here the main point to notice is that each edge is responsible for creating only two triangles. Actually the mesh describes the surface in 3D.

Points worth remembering:
3D models are not an extension of 2D images. These are the collections of triangular meshes in 3D space.

14.5.2   Embedding Domain in 3D Models

The embedding domain depends on the cover object where we are going to embed the watermark. In the case of 2D images, we had two domains: spatial and frequency. In the spatial domain we directly modify the pixel values whereas in the frequency domain, we modify the frequency coefficients of the 2D image. In the case of 3D models, we have more degrees of freedom and we have more features to modify such as vertex positions, edge connections between vertices (also called topology), properties of the surface, i.e., vertex color, surface color, etc. By this way we have many options for embedding. But here the two most common domains used for embedding the watermark into 3D models are geometric features and topological features.

14.5.2.1   Geometric features

As we know that 3D models are nothing but the collection of a number of vertices. Each vertex in space has some 3D coordinate. With geometric features, we mainly focus on the position of the vertices. A watermark can be embedded into the 3D model by just modifying the vertex normals or positions of the vertices. Vertex normals are directly related to the curvature of the surface. Changing the position of the vertex means changing the coordinates. One can make an algorithm that modifies the coordinates as per the watermark bits and these changes must be reversible in nature so that at the receiver end, the receiver can judge the watermark bits.

14.5.2.2   Topological features

Here we deal with the topology of the meshes. Topology means the connection of edges with vertices. In order to embed the watermark bits, one can reassign the edge to new vertices. The internal structure of meshes may vary without affecting the actual shape of the 3D model. During reassignment of the edges to vertices, we have to make sure that each edge must contribute to only two triangular meshes. We can see in Figure 14.10 how the topology features are modified.

Points worth remembering:
In 3D models, mainly two features viz geometric and topological are used to embed the watermark.

14.6   3D STEGANOGRAPHY AND SECRET SHARING

In the previous section, we obtained a basic knowledge of 3D objects. 3D steganography can be developed just like 3D watermarking method. Here we need to choose some random locations on a mesh model where we can make some changes geometrically or topologically in order to embed the 0 or 1 bit of the secret message. One

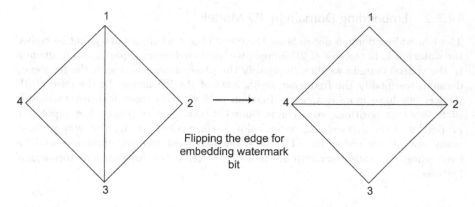

FIGURE 14.10 Example of topology feature modification.

can choose the color and shading features of the model but they are more vulnerable to simple intentional or even non-intentional attacks. Hence we need to develop an algorithm which is more robust to these attacks.

3D secret sharing is an emerging area of research in which just like 2D secret sharing, we need to share 3D models with random or meaningful shares. These shares will also be 3D in nature. Suppose we need to share a 3D model with two random shares. Let us consider a vertex V_1 of this model having coordinate values x_1, y_1, z_1. If we are dealing with geometric features, then we need to find two random coordinates x_2, y_2, z_2 and x_3, y_3, z_3 in such a way that

$$x_1, y_1, z_1 = Share\ Generation\ Function(x_2, y_2, z_2, x_3, y_3, z_3) \qquad (14.1)$$

where the *Share Generation Function* is any developed function which splits a co-ordinate into two random coordinates and then merges them to find out the original one. Readers can pursue in this direction to develop their own algorithms. In order to deal with 3D models, readers can work with MeshLab with MATLAB. There are many more open source tools available for handling 3D models.

Points worth remembering:
3D models can be handled practically with combination of MeshLab and MATLAB.

14.7 BIOMETRICS AND TELEMEDICINE

Biometric and telemedicine are the two latest areas in which we can apply image-based security approaches to enhance security features and efficiency. We know that nowadays, a great deal of authentication is carried out only by biometric data either by fingerprint or face or iris, etc. Therefore we can understand the importance of this biometric data. This data is responsible for proving our identity in many places. We can protect it by digital watermarking techniques. Telemedicine provides health care access to rural and remote locations by enabling practitioners to evaluate, diagnose and treat patients remotely using the latest telecommunications technology. It facilitates patients receiving expert medical suggestions without having to travel. Rural health care practitioners can use telemedicine products to capture and transmit medical data and images. So in the case of telemedicine, we see that a secret medical image is being transmitted from patient to medical expert and this requires security. Hence here we can secure the transmission of medical images using a multitone visual cryptography method. Here we stipulate multitone VC because in cases of medical data, it is very difficult to explain the smallest details of a medical image using only two intensities (halftone VC). Securing medical data during the transmission is very important because even a little modification in a medical images may lead to wrong diagnosis.

Points worth remembering:
Biometric and telemedicine are two emerging areas where image-based security techniques can be applied. Biometric data can be secured with watermarking whereas telemedicine data can be secured with secret sharing or VC.

SUMMARY

- Image-based security techniques can also be used to protect *other multimedia objects like audio, video, etc.*

- In case of *audio watermarking* we have the freedom to choose the watermark format (either audio or image).

- An images is a two-dimensional matrix whereas audio or speech is *one-dimensional vector.*

- Readers can choose their own recording format for speech and any copyright logo (image) or speech signal for a watermark.

- Fragile watermark bits are generated by applying any self-embedding approach on the bits of the audio samples and embedding into the LSBs of sample.

- Shamir's basis matrix can be applied on the binary stream of speech to develop the basic *audio secret sharing approach.*

- Image-based security techniques are applied on each frame of video and finally all frames are combined together in order to achieve *video steganography/watermarking/secret sharing.*

- *Random selection of frames* of video for watermark or secret message embedding can enhance security.

- *3D models* are not an extension of 2D images. These are the collections of triangular meshes in 3D space.

- In 3D models, mainly two features viz *geometric and topological* are used to embed the watermark.

- 3D models can be handled in practice with a combination of *MeshLab and MatLab.*

- *Biometric and telemedicine* are two emerging areas where image-based security techniques can be used. Biometric data can be secured with watermarking whereas telemedicine data can be secured with secret sharing or VC.

Index